FSS

FIELD SECURITY SECTION

BOB STEERS

ISBN 0 9528129 0 8

First published in 1996 by Robin Steers, Eastlays, Hailsham Road, Heathfield, East Sussex TN21 8ES.

Distributed by the Intelligence Corps Association, Templer Barracks, Ashford, Kent TN23 3HH.

Printed by Olivers Printers Limited, Cesar House, Eastwood Road, Bexhill-on-Sea, East Sussex TN39 3PT.

Dedicated to the past, present
and future members of the
INTELLIGENCE CORPS

Contributors

The narratives appear in the numerical order of the Section numbers

The Introduction

"In many ways Intelligence is more dependent than other arms on the quality of its personnel."
Brigadier E. Mockler-Ferryman

'Ye shall hear of wars and rumours of wars'. If anything this is a good enough reason for the existence of the Intelligence Corps for, indeed, its motto is 'Knowledge gives strength to the arm'. Our raison d'être is to keep our commanders well-informed of what could confront them and prevent unfriendly forces from obtaining the corresponding information.

But this is not a history . . . that has already admirably been written by Anthony Clayton. This is an adjunct to his work which, hopefully, will give the reader an insight as to the kind of person the Field Security Section attracted during and after World War II and something, within the constraints of the Official Secrets Act, of the work which was carried out.

The concept of the Field Security Section lasted for some 20 years, i.e. 1940 to about 1960 and was just one of the activities of the Intelligence Corps. In many ways it was at the sharp end of the military intelligence organisation. Wartime members often claim that the motto should have been, 'First in, last out' as, frequently, they were among the first to enter a captured town in order to secure vital buildings and arrest their opposite enemy numbers. This is illustrated by one section which arrived in a Dutch town as the Gestapo were still loading their vehicles before attempting an escape. Again, FS were the first troops into Salerno, Italy on 9th September 1943 and, having secured the post office, intercepted a telephone call from an Italian Colonel who was in German-occupied Naples. They assured him that the British had landed and, through questioning, discovered that he was calling from a brothel. Luck also played its part as when a FSNCO, who was a philatelist, went into a small post office in the Western Desert to buy a set of stamps only to find the ration strengths of the German Afrika Korps facing that particular front.

FS operatives made contact and worked closely with underground and resistance bodies in the invaded continental countries and were successful in hunting down and arresting large numbers of enemy agents. Only occasionally was there drama. Most of the work was governed by sheer commmon sense and the understanding of the thought processes of other nationalities. A good example of this is that on entering a town in Malaya, a Section would raid the photographers' shops. This was because they knew the Japanese had . . . and still have . . . a predilection for having their photographs taken. Thus, the Section would easily acquire the likeness, names and addresses of the local Kempetai, the Japanese Secret Police. As the Cold War developed Sections found their role changing. In Europe they were working with friendly intelligence organisations in the Counter Intelligence mode and were increasingly becoming interested in industrial and military developments in the eastern European nations, even stretching into the then USSR. In the Far East they were actively running line-crossers during the Korean War and involved in counter-insurgency in Malaya and the Middle East.

The principles of Intelligence and Counter Intelligence have not altered much since the year 1451 BC when Joshua sent two line-crossers into Jericho to ascertain the strength of the defences and the state of the morale of the inhabitants . . . information any modern commander would require. They were assisted by a resident prostitute named Rahab, thus, a member of the oldest profession helping two members of the second oldest profession. Even in this day of the micro-chip, success is often dependent upon face-to-face contact and the understanding of one's fellow being.

During World War II FS attracted journalists, schoolmasters, actors, lawyers and self-starters . . . just the types to blend into the work which was required of them. After the war we had the cream of the National Service intake to complement the regular backbone of the Corps. Recruiting methods were sometimes unorthodox, such as the Army Council Instruction which stated, 'ability to play bridge an advantage'. In this respect the reader will arrive at his or her own picture as to what constituted the FS man. Ability to play bridge? Yes, the section heading for the Normandy beaches on D-Day passed the crossing by playing a rubber or two of bridge . . . probably unique amongst the odd million chaps who went to France.

The low-key contributions will indicate to the reader some of the FS characteristics. Wit, humour, humanity and a stubborn professional attention to detail; the ability to cope with a fanatical SS Nazi official, communicate with head-hunters, diplomatically contend with the illegal Jewish emigrants attempting to enter Palestine, and, en passant, lunch with the King of Italy.

Nearly 45 years ago a colleague commented, "We are really a Church Militant," and how true it was.

To assist us in our work there was the Spot Card. This was so named because of the inch diameter black spot on the cover. Page two had a photograph of the holder with description and signature. On page three was written in English and various languages, depending on which part of the globe you were in, but, curiously, never in Russian, the following text: "The bearer of this pass has authority to be in any place, at any time, in any dress and all authorities are called upon to assist him in the execution of his duties". And it was signed by a very senior person. This, clearly, was a carte blanche but woe betide anyone who abused its legitimate use.

One of the writers succinctly described our raison d'être. "Nobody really quite understood what we were doing – we enjoyed a certain mystique and even senior officers were chary of challenging our activities". Quite right, as any old hand will know. Get yourself a good billet such as a palazzo or a decent hotel, a good cook, set up office and then start making contacts. Contacts are the basis of most Intelligence work.

Occasionally, we were misunderstood. As instance:

1942

Mobilisation Officer (not Corps): 'Well, what languages do you speak?'

"I have Arabic, Greek and Old Greek, I can also read cuneiform and heiroglyphics".

Mobilisation Officer: "What are hieroglyphics?"

"The language of the Pharoahs, Sir".

T.G. Bibby, formerly of 319 FSS, Faroe Islands.

This, then, is a picture of the FS man and, hopefully, the ethos of the Field Security Section. Many writers stressed that after the passage of half a century some of the minutiae may not be as accurate as their memories would wish. The final contribution by the late W. Sedwick-Rough could easily be entitled *The Conclusion*, as he summarises much of what has gone before.

This book was compiled at the behest of the Intelligence Centre, home of the Intelligence Corps, not to state, "this is what we did" but rather to say to later generations of Intelligencers, "these were our mistakes and our sucesses, now it's over to you".

In conclusion, my thanks to all those who wrote to me and my apologies to those whose articles have not been included. This was mainly done to avoid duplication of events and locations. Everything which I received, however, has been passed to the Corps Museum and now rests in the archives there.

I also wish to express my thanks to Mrs Jan Carpenter of the Intelligence Corps Museum for her enthusiastic assistance, to Major Dick Shaw for his occasional touch on the tiller which was appreciated and, of course, to all contributors . . . what follows is the sum of THEIR experiences.

Heathfield

Bob Steers
March 1996

3 FSS HQ 1 Brit. Inf. Div.

David Scarisbrick
The OC of this Section for 3½ years

I was really rather proud to be commanding 3FSS and it was good to be with a very senior regular Division. In my own mind I am quite satisfied that there was nothing in my security 'career' of any interest to anybody but myself and I know that I really achieved precious little . . . but I had a marvellous time!

I share all the memories of the early recruits at Mytchett and Sheerness. The pink-cheeked young lance-corporal in charge of our barrack room at Mytchett was Dennis Bloodworth, later Far East correspondent for 'The Observer' (and I later took over from him in Tunisia, somewhere near Madjez-el Bab, when 1 Div. took over from 78 Div.). One of the shining lights at Sheerness was a smart and soldierly Malcolm Muggeridge who gave a brilliant lecture on 'Propaganda'. We were at Sheerness at the time of Dunkirk and we had the job of mustering and disarming French and Belgian troops who sailed into Sheerness.

It was a young Lieutenant William Sedgwick-Rough who led us out of Sheerness as his very first command (I forget the section number). He was soon succeeded by an older Captain Albrecht and, soon, a detachment of five under my command went to join No. 6 Commando, based on Dumbarton and waiting to go overseas on some operation or other. We were trained to Commando standards on all infantry weapons (our OWN section Tommy-gun!) by a young Sergeant Roger Delgado, formerly of the Leicester Repertory Theatre and later 'The Master' in the 'Dr. Who' television series. A hint was dropped that Portuguese would be the appropriate language on this operation . . . Azores or Cape Verde Islands? We were a French-Spanish section but rusty on Portuguese and I recall that we scattered around Glasgow's second-hand bookshops to pick up helpful books, being deliberately casual!

After Sandhurst and the usual course at Matlock(memories of Evelyn Waugh on the War Intelligence Course) I took over 3 FSS at HQ 1 Division. To Algiers for the Tunisian Campaign. Of particular interest to me were the happenings at the end of that Campaign.

A substantial part of the German forces retreated into the Cap Bon peninsular and surrendered there. Many of their weapons they handed over to the local Arab population, asking them to take care of them, because "We shall return!". My Division settled in the Cap Bon peninsular and I became responsible for security in that territory.

We were faced with a second problem. Some scores of the more intrepid German soldiers, mainly NCOs and warrant Officers, did not surrender but 'took to the hills' and the remote peninsular – still armed, of course – and persuaded (or forced) the simple villagers to shelter them. They hoped, in due course, to slip away westward to neutral territory . . . and, perhaps, some did.

Two problems, therefore, for me in 'my' territory – a vast quantity (as it turned out) of weapons in the hands of the local population, and an unknown number of resourceful German soldiers in hiding somewhere in our midst.

The French Colonial Service officers had moved back into 'their' territory and began – with my cooperation, I fear – a programme of rather heavy handed raids and searches. With regard to weapons there was little success and only a handful of German 'fugitives' gave themselves up. Then, in a change of policy, a new enlightened and very sophisticated 'Caid' (a senior Arab notable) was appointed to the territory. He asked for 14 days to solve the problem. He toured the villages, talked to the Sheiks and the people and within a week some 17,000 weapons had been brought in to the gendarmerie stations and many German fugitives were forced to give themselves up. (There were one or two violent episodes.) I recall that one of my chaps –

Charlie Hale, a quiet schoolmaster – was roused from sleep by excited villagers and told that two armed Germans were sleeping in a nearby cave. Poor Charlie crept along, pistol at the ready, found the sleeping men (with a Schmeisser machine-pistol to hand), resourcefully yelled, "HAENDE·HOCH" and bagged his men. Before movinq to Italy, my division took part in the massive combined operation (that I never hear mentioned now!) to capture the Italian islands (between the toe of Sicily and Tunisia) of Pantelleria – a thorn in our side because of the great air-base – Linosa and Lampedusa. A dress rehearsal for subsequent sea-borne operations.

In Italy, my division was singled out, with an (initially) small American force, for the Anzio beach-head operation south of Rome, intended to draw German forces from their stubborn resistance at Montecassino. A scaled-down version of my section went with the Division. My main 'security' recollection was the following: Complete surprise was essential for the success of the landing. Absolute secrecy was vital. Against this background, the Divisional Commander held a final briefing conference in the divisional assembly area around Castellamare for commanding officers and their adjutants and 2ICs. Precise plans were spelt out. No note-taking was permitted. But a young adjutant DID take notes in his Field Service Notebook with enough detail to 'identify place and time!'. After the journey back to his regiment (a tank unit) in the back of a truck, he realised to his horror, that his notebook had slipped out of his map-case and out of the truck, on the road from Castellamare. The alarm was raised; a quick search along the road (which passed through populated areas) yielded no result. The Divisional Commander told the Operation Commander, Mark Clark, who said the operation must proceed, regardless. The Field Security Section was called in! That evening, until long after nightfall, we searched and called at houses and farms along the route, discreetly questioning. At daybreak we started again, then noticed a Pioneer Section already working on road repair. A soldier was asked, "Were you on this road yesterday? Any knowledge of a notebook dropped on the road?". Oh, yes, and it was handed, unopened, to their officer. Quickly along to the young officer. Yes, here it is, unopened, in my strong-box. Quick journey back to Div HQ. Immense relief. A successful 'routine' security job. It was known, incidentally, that there were plenty of rather low-grade Italian agents operating (with standard radio transmitters) in all of the territories that the Germans had withdrawn from, so that the possibility of information being transmitted to the enemy could never be discounted and quite elaborate deception measures were regularly being taken to confuse the enemy regarding reinforcements, reliefs, etc. So the risk about the Anzio plans was a real one.

Harking back to Cap Bon peninsular in Tunisia, SOMEBODY must have thought our efforts worthwhile, because the French Military Governor prevailed upon the Bey of Tunis to appoint me to the Order of Nichan Iftikher of Tunis. This is a source of great amusement to my family.

The five months of the Anzio beachhead battle were ferocious but once most of the civilian population had been shipped out, we had relatively little to do, except keep our heads down.

It must have been 4th June 1944 and I was in the billet of an American Counter Intelligence Unit (FSS equivalent) some miles south of Rome. The great battles of Anzio and Cassino were over and we were hourly expecting news that the pursuing Allied Forces had entered the city. With us was Prince Giorgio Borghese who had stayed behind in the Anzio beachhead to care for the few remaining civilians and who was desperate to see his family in Rome. Suddenly, somebody burst in, "We've entered Rome. Jump in my Jeep, Giorgio, and let's go!". I went too. We streaked past moving columns of American armoured vehicles. On the outskirts of the city some sporadic firing in the distant checked us and we turned off along a side-road. It was too quiet for my liking. Suddenly, a figure leapt out into our path, waving a flag and proclaiming himself a 'Resistance fighter'. "There's a small German rearguard detachment just round the corner and one or two armoured vehicles further on. God Speed!". A rapid U-turn. It later became clear we had entered Rome one day too soon. As we drove back we met the first American tanks, a young officer cautiously leading on foot, pistol in hand. We told him what we knew. Back at the junction with the main road into Rome an American tank colonel was conferring with his colleagues whilst sporadic firing still continued beyond the final viaduct into the city. "Hey, what's it like up there?". We gave our report and he directed more tanks up "our way". Later that day Rome was indeed liberated and at dawn my Sergeant Major and I were in St. Peter's Square.

All was silent. Exhausted American troops lay sleeping on the pavements. We walked across a silent, deserted St. Peter's Square and up the steps of the Bassilica. As we approached, a door opened, and a smartly dressed, Panama-hatted gentleman emerged from St. Peter's, hand outstretched – "Welcome, my name's Montgomery!". He was the British Envoy to the Vatican; he had been immured in the Vatican City since Italy's entry into the war; this was his first emergence and we were the first British soldiers he had seen. A moving and memorable moment.

After Anzio, the Division fought its way up Italy, via Florence, and finally came to a halt in the snows of the Appenines, when the entire Allied advance came to a stop at the Gothic Line. The Division was depleted and exhausted and before long was evacuated to the Middle East, to Palestine. A marvellous, busy year, leading to demob.

I loved my years as Field Security Officer to a regular Infantry Division and as a Commander of a Field Security Section. My independence was almost total. Nobody really quite understood what we were doing . . . we enjoyed a certain 'mystique' . . . and even senior officers were chary of challenging or questioning our activities. We could get away with almost anything!

Our lips were sealed! I recall that when, after the conclusion of the Tunisian battle, I approached our Divisional G2 (the wonderful Lt-Col (later Brigadier) Gerald Thubron, I think) and suggested it was time I was promoted to Captain, he simply said, "Of course, you've worked well for us. What must I do? Right, write your own recommendation and I'll do the rest."

And that was that!

3 FSS

Malcolm White

Anzio – 22 January 1944

On the deck of the Tank Landing Craft we were woken about four o'clock by a horrendous explosion – a whole battery of rockets from a ship somewhere to port had been fired at the shore. We waited, it seemed for hours for the forward troops to clean up beaches and get inland.

Eventually I and my BSA motor-bike left the TLC and went gingerly up the rolled-out metal roadway and over the sand hills. There had been no response from the Germans to our rocket challenge and now no sound of small arms or artillery.

I presume that I made contact with the 2nd Infantry Brigade HQ to whom I was attached for the Anzio operation and off I went north along the coast road to find my 'objective'. After 50 years I can't remember what my 'objective' was – a radio station or something on the coast.

I was chuntering along the road which was slightly raised above the surrounding flat countryside. No sign of anyone English, German or Italian when suddenly – "Oi stop!" – an English voice from a ditch on the right. I stopped, turned and went to see my challenger. 'Do you realise we are the front line? If you hadn't heard me you would have run straight into German lines.'

Not wanting to become a POW. I thanked him profusely and made my way back towards Brigade HQ. My 'objective' would have to wait for another day. With no signs of German military I did not think it would be long before I could reach it.

In the build-up of the battle for Anzio our main task was shepherding and screening civilians away from the forward areas. The Germans had brought down troops (from Norway?) and were determined to push us into the sea, throwing a lot of stuff at us.

3 FSS had a farmhouse in an ammunition dump near the beach and once the main battle had been fought and we were still in Anzio and not adrift behind the German lines or POWs in Germany, improving our Italian and daily visits to Division and Brigade HQs became routine. One morning Bill Mather and I went through the ammo dump to Div HQ and, passing a stack of boxes, we were astonished to read on them, POISON GAS. There wasn't supposed to be poison gas nearer than Casablanca. We reported this 'oddity' to Division. The next time (? day) Bill and I once more went through the ammo dump the stack of boxes marked POISON GAS was nowhere to be seen.

Firenze – Summer 1944

We were screening refugees as they came from or through Florence and over the Arno. We had set up our roadblock in the hills south of the Arno, I think it was at Grassina; a useful but unspectacular job. We were approached by an officer in our artillery or infantry. They were having difficulty in identifying some German mortar units who were lobbing bombs at his lines from behind hills just north of Florence. Could we help? It so happened that an Italian golf coach from the Florence Golf Club had a shortwhile back introduced himself and asked if he could help. He had some English and saw no difficulty in obtaining the information the officer wanted. Off he went down towards Bagno a Ripoli, crossed though our lines, over the Arno and returned a little later with the information required.

3 FSS was always ready to oblige!

Fiesole 1944

I reckon I was very fortunate. When the Germans had fallen back towards the Gothic Line, I was told to set up shop in Fiesole. So I fould a small office on the town square and set about making the usual contacts with the police and other useful agencies. In the short time available there appeared to be one person who was suspect: pro-German or Fascist. He was a well-established artist(?), middle-aged and well in with the Church. He had painted murals in the Franciscan monastery in Fiesole and soon I had priests and others kneeling before me in my office pleading for him. I was able to see one of the leading Italian conductors, Signor Gui, who was non-commital about him. Time being very short, I decided to send the suspect artist (whose name I cannot remember) back to Corps Field Security for further questioning. I was, thus, obliged to face the howls of despair and anguish from the local representatives of the Church of Rome.

Reminiscences of 3 FS Section NCO 1945–65

P.M.S. Ward

Amongst my Libyan friends was one Haj Ramadan; a friendly person with a keen sense of humour and a genuine Anglophile. He was also a member of the Upper House of the Libyan Parliament and, like me, he had a taste for the local palm wine or 'Leghbi'.

There was no finer leghbi than that of the oasis of Zanzur, some 12km west of Tripoli. We used to meet at a little farm at Zanzur for a drink and sometimes the farmer's wife would prepare for us a succulent dish of cous-cous. Years later I used to visit this same farmer with Sergeant (later Warrant Officer 2 in BAOR) Dave Varley and his family. I will always remember the look of amazement on the little Libyan children's faces when we all visited Dave's quarters in Tripoli for a return visit and a meal of fish and chips!

At these meetings in Zanzur there were always a few local Tripolitanians and sometimes a few members of the FLN (Front Liberation National) from Algeria, who were in Libya to publicise their cause and obtain vital supplies for their struggle for independence. Once I was able to help the FLN obtain a quantity of urgently needed antibiotics and other drugs essential for the treatment of their wounded being nursed in hideouts in the mountain areas of Algeria. If you knew the Libyan border area – and I knew it well in those days – it was a smuggler's paradise. To prove my point I took my then OC, Captain Roy Bignell, in my DKW car and we were able to circumnavigate the frontier post by following the well trodden camel tracks via Reheibat.

One evening Haj Ramadan remarked to me that it was a pity that the English were not able to spread their news and views throughout the country as some nations did. He confided to me that a friend of his, a merchant, Haj Mohamed, had received from the Russians a supply of magic lanterns which, when lit in even the remotest Bedouin tent, would pour forth the propaganda of Radio Moscow. A few days later I was able to track down the merchant Haj Mohamed, who regretted that all his supply of magic lamps had been distributed, but that there was a further supply to be collected from a Russian merchant ship lying in Tripoli harbour. Next evening Haj Mohamed and I, mounted on my old 1918 BSA combination, entered the port and, after crossing a few palms with silver and taking a sip or two of vodka, the deal was done. The Russians were intrigued with my old 1918 BSA; especially the acetylene lights which the Captain wanted to buy. My wife and I had come by road to Tripoli from Bradford in this machine and I think it was the oldest vehicle for which a 1771 claim for motor mileage allowance had ever been made. The Command Secretariat were not amused !

Arriving back home in Gargaresc I unpacked the crate with the magic lamp and found a conventional roadcast radio receiver, a control box with a meter and a mallory type vibrator, and a standard type paraffin lamp which had a circular finned device mounted at the top of the lamp glass. This device was a thermopile which, when heated by the lamp, produced potential of about two volts. The voltage was transformed by the control box to the necessary supply voltage for the radio. This was of course in the days before transistors and caused a sensation amongst the Libyans who saw the set in action.

Although it was new to Libya, my research has found that a similar device was marketed by the Gas Company at Newcastle on Tyne in about 1928.

My FS Section was interested in the radio device and the Embassy was very interested, but I had to recover over my expenses – so I sold it to the American OSI Wheelus Air Base for £20. Looking back, I regret that I could not afford to keep it. Perhaps it might even have ended up in the Corps Museum at Ashford . . .

5 FSP

Anthony Howard

In October 1939, I enlisted in the FSP after hearing an announcement on the BBC. I went to Mytchett, but can remember little about the training except a lecture on how old ladies in occupied France in the first war, recorded details of German troop trains in their knitting. In those days I was too respectful of authority to ask what relevance this information might have to our forthcoming duties with the BEF. But of much greater value to me proved to be the parting words of the Office Corporal. "A word of advice, Howard, don't be too intelligent." My one bit of fortune was to meet the late Hugh Grant, a professional actor, who was to share many of my vicissitudes in the years to come.

In France we eventually ended up in Lille as part of 5 Section (4 Div I think), under command of the delightful Sir Basil Bartlett. Our regular Army Minder was a Sergeant Baird, who did his unsuccessful best to keep us in line. I remember Basil commenting "I had hoped that Sergeant Baird would turn you into soldiers, but you have succeeded in turning him into a civilian!"

I was posted to Halluin on the Belgian frontier, where I had the experience of catching my only spy. I was told that there was a suspicious character who frequented a local cafe, wore British uniform, spoke French and took notes. This assignment presented no difficulty as I already lunched there every day – to keep the frontier post under observation should anybody ask. However, the suspect never appeared and I assumed that he had moved on. Until, that is, I was relieved by two men from another section. I introduced them to my various contacts, ending with the proprietor of my café. I asked him if he knew anything about the mysterious spy. "C'etait vous monsieur!" And so it proved. Not only did I speak fluent French, but I always took a French text to study over lunch and made notes of the difficult phrases. Thanks to that episode, my wife has been able to tell our children and, later, grandchildren the story of "How Daddy/Grandpa caught a spy."

Early in 1940, it was decided to send a FS Section of French and German speakers to Metz, the HQ of Saarforce, which was set up to provide patrolling experience in front of the Maginot Line for relays of BEF troops.

Our regular duties were undemanding, divided between looking out for suspicious characters at the railhead when troops came and left, and going through the barracks and billets for compromising documents such as marked maps used as lampshades.

Occasionally more unusual assignments came my way. There had been a report of the odd parachutist landing in a field above the village where we were billeted and – on the theory that this might develop into a nightly event. I was sent to keep observation, though it was never made clear what I was expected to do should Heinrich appear. I was lying hidden on the long grass when I heard footsteps approaching. Raising my head, I was challenged by what turned out to be an elderly peasant armed with a shotgun – the local Dad's Army perhaps. A mutual challenge having resulted in deadlock, we escorted each other to the local mayor, who fortunately recognised me as I was billeted with his sisters. Looking back, it seems quite extraordinary that we were never provided with any means of identification that we could produce to the French civil or military authorities.

Time passed pleasantly enough with regular visits to the police, the local press and the 2ième Bureau – this last always in a café. And when all this got too hectic there was always the cinema and excellent restaurants. It was as if we were in a time-warp, because this routine continued until well after Dunkirk, which might as well have been on another planet for all the effect it had on those in charge.

But all that was to change, when – shortly before the French surrender – we were ordered out in a hurry and put on a troop train which trundled slowly southsouth until the drivers abandoned it near Vesoul. We then took to the road, ending up in Besancon, where with about fifty compatriots we joined several thousand French POWs. At this point my career was to change from that of Field Security Policeman to becoming a Stalag Interpreter for the next five years.

In December the barracks were evacuated, leaving only a few French, Hugh Grant and myself. What happened next was the influx at all hours of night and day of some three thousand British women (including many nuns) and children, who were living in France and were now to be interned in very rugged conditions. It fell to Hugh Grant and myself to help them to get organised. Within the first couple of days we were asked to arbitrate between the women of British origin, who insisted on having the windows open at night, despite the bitter cold, and those who were originally French, but had married British men, and wanted windows to stay firmly shut. Solidarity in face of the common enemy was restored by having separate barrack rooms for the French. Within a few days an effective organisation was up and running, including classes for the children, lectures and concert parties. The nuns provided teachers and nurses; in the latter capacity they were unsurprisingly preferred to those from the German Red Cross.

While all this was going on, the German Camp Commandant, an inoffensive sports dealer, kept well out of the way, emerging occasionally to drive round the camp in his Volkswagen. On one such occasion he ordered me into the car, only to ask me "Howard, what have we done that God should punish us with all these women?".

Towards the end of January 1940 this very rewarding period of our captivity came to a sudden end when Hugh and I were suddenly removed from the camp and taken to Germany, where we ended up in a military prison in Freiburg, on suspicion of espionage. How, they asked, did we come to be in a civilian camp? There seemed to be an assumption that we must have come there by devious means, since no efficient German military administration would have countenanced such an unconventional arrangement! They were unconvinced by our explanation as to how we came to be there. While they were sorting this out, we were confined to the upper tier of the prison, where I learned from the occupant of the next cell that he was a deserter from the German army and serving a suspended death sentence, as were all the other occupants of my floor. This information did not exactly improve my morale, especially when the unfortunate fellow raised his voice to the prison officer and was promptly shot on the spot. To my relief, the officer was removed to other duties.

After three anxious weeks we were cleared and returned to a Stalag. I spent the rest of my captivity as an interpreter. At times I took it on myself to filter some of the exchanges between the British and German camp commanders and although over time, they must have understood each other without my help, they chose to accept my less charged translation.

Postscript

In May my wife and I visited Besancon to renew acquaintance with the French woman who had 'adopted' me when I was a POW there. She sent me parcels until I went to Germany, where she used to write to me. Alice Grelier and I have kept in touch over the past fifty-five years and we are agreed that we owe a debt of gratitude to Hitler for bringing us together!

8 FSS

Derek South

In March 1946 I was entrusted with my first solo arrest assignment. I was driven 'downtown' from our elegant quarters in Essen Bredeney, the home of a former Krupps director, by Franz, who handled the Hansa saloon with a panache and flamboyance which apparently enabled him to disregard the condition of the cobbled, bomb-pitted roads of the Rhur.

I was delivered safely however to the block of grey, heavily damaged flats which housed my quarry, a former SS man. In terms of rank, of no great status, a Scharführer I believe, but in physical terms, imposing. (A subsequent check revealed height at 6'6".)

I climbed the winding staircase in virtual darkness but managed to locate the right flat. My 'target' himself opened the door and the slight improvement in light from within created a formidable silhouette. I followed him into the room and noted that it boasted a single, cracked window in one corner, overlooking a small hand basin. He turned towards me and I realised that his face was half covered with shaving soap and in his right hand he held an old fashioned cut-throat razor. I informed him that he was to be questioned at our HQ and then as a former SS, he would be transferred to 4 CIC internment camp at Recklinghausen. Quite gently he asked if he could complete his ablutions. I pulled back a couple of paces and told him to get a move on. As he resumed at the washbasin the only light penetrating the gloom seemed to radiate from the moving blade. Today, we talk of insecurity, I remember the feeling! I unclipped the holster flap on my Webley pistol and allowed an increasingly sweaty palm on the butt as I followed the hypnotic movement of the blade in his hand.

If such a scene were enacted on a stage I would scorn the cheap drama but at the time it proved difficult to take a detached view. I waited in complete silence until he had completed the shave. He crossed the room and began to pack a small case of personal items, each of which I sanctioned. The extreme docility of the towering figure remained a feature for which I was quietly grateful. Lodged in the rear of the car, we were briskly returned to HQ and the case ended with his transfer to internment the following day.

This kind of reaction to arrest became very familiar among former SS and SA but less so among SD and Gestapo personnel who seemed to regain confidence in the knowledge they had accumulated with regard to communist organisation and methods of subversion which they sensed were becoming objects of increasing interest as far as we were concerned.

There were some FS who were motivated, quite understandably, by a desire for retribution. It was very much a matter for the individual. In my own case I had entered Germany with a willingness to hate. My elder brother had been killed shortly before D Day flying with the RAF and our family home in Kent had been destroyed by a V1 flying bomb one month later. Temptation for revenge was real enough. Whilst participating in the breakdown of the Nazi movement proved immensely satisfying I was not able to find any capacity within me for durable hatred.

The utter devastation which I witnessed in the Rhur and later in Hamburg and Berlin convinced me that for most of the ordinary population a heavy and appropriate price had been exacted in retribution for their blind devotion to the evil of National Socialism.

The price for the sins of the Nazi leadership had yet to be announced and this also provided an interesting occasion for me after I had been transferred during 1946 to another section.

Autumn 1946

Transfer from the Rhur to Wesphalia represented a change in the political 'colouring'. In the Rhur there had been a lively struggle for power between the SPD and the KPD. As Communist support declined (and was discouraged – and eventually banned) the SPD assumed dominance. In Westphalia the CDU predominated although the SPD remained active within the small townships. The first post-war local elections were to take place in the autumn as though concluding the chapter on Nazism before opening that of a democratic Germany the Nuremburg verdicts were announced.

I happened to be within a German household pursuing some political enquiries when the Deutsch Rundfunk interrupted a background programme of music with the announcement from Nuremberg.

The dramatic impact of the sentences was powerful, compelling rapt silence within the listening group.

After each familiar, and for me, notorious name was announced pedantically uttered 'Zum Tode durch den Strang verurteilt' (Sentenced to death by hanging) burned into one's consciousness. As a member of the Occupation Forces I felt complete satisfaction in the statement, happily accepting the inclusion of the military Jodl and Keitel with the political criminals of the Third Reich.

For the others however, reaction appeared to be one of stunned disbelief. Not arising, as far as I could see, from any sense of injustice but from the shock that former German statesmen and top brass who had featured on the world stage should now be regarded as candidates for the hangman. The silence continued and I realised that the presence of a British uniform was not helping them to come to terms with the verdicts. I simply shook hands without further comment and departed.

Whilst there was a considerable decline in simple anti-Nazi activity for us in the latter half of '46 I do recall one other unusual case of arrest which remains vividly in the memory.

Together with CSM Jim Palmer, I entered a crowded dance hall one winter evening having received the informant's tip-off concerning the likely presence at the dance of a wanted SS officer. On this occasion, a higher ranking Obersturmführer, although reportedly in his early twenties he was typical of such rankings gained in the closing stages of the 1945 fighting.

At this point I must praise Jim for his excellent service as CSM. Possessed of a photographic memory. An astonishingly accurate eye for detail and his retention of names and records was quite phenomenal. Also a great friend. Jim was an ex-Para. On some occasions when coming directly into contact with the native Germans however he was, to put it mildly, inclined to become vitriolic. More Para shall we say than I. Corps?

On the occasion in question, after entering the hall, I stayed at the lone exit door to prevent any early departures. Jim proceeded to storm through the crowded dance floor, thrusting couples aside and then leaping on to the low stage to bring a halt to the pounding Teutonic 'beat'. Our CSM swung round and facing the now sullen and resentful dancers he bellowed in execrable German (akin to Churchillian French) announcing that all ID's were to be checked at the door. After all had been checked the dance could resume.

As on many previous occasions, my Webley holster was unclipped again. Not that any dancer noticed. They were stunned by the incredible performance and mutually formed into a queue before the small table I had set up. The atmosphere was tense however and perspiration flourished, having little to do with the hall temperature. With aggression in the air Jim returned to me and then together we checked each Ausweiskarte. Approximately halfway through the queue we were delighted to discover our quarry. Much to his surprise and obvious consternation we ordered him outside to the waiting Mercedes. No interest was shown by the German crowd in the possible fate of their compatriot as we waved a resumption of the dancing.

I bundled our man into the rear seat and joined him, with hand still comfortably in touch with the pistol butt. Jim meanwhile vented his undiminished rage on the accelerator pedal being further infuriated by the reaction of the youthful ex-SS officer which was immediate and surprising.

He turned to me tugging at my sleeve and pleading to be released. It would, he claimed, be devastating for his mother. The news might kill her. She must in any case be told where he was and what had happened to him.

The appeals reached a crescendo and sobbing loudly he made to lean on my shoulder for which he received a swift and violent rebuff.

The journey was completed with our detainee huddled miserably in the corner of the car.

We offloaded him to the German police at Werl Prison where so many of the former Nazi hierarchy had been lodged before him. As we entered the gaol offices, our SS prodigy straightened up, unbelievably suffering an apparent retrogressive time warp, for he clicked his heels and was about to raise his arm in the Nazi salute when one of the police seized him violently, swung him round and hurled him through an adjacent cell door.

We completed the documentation with the conviction that our Obersturmführer was to be indoctrinated into democratic ways in rather undemocratic fashion.

Jim appeared to be very satisfied with the evening's work.

8 FSS

Derck South

Essen, Germany – Spring 1946

There were a number of senior Int. Corps personnel whose linguistic ability had been relevant in previous 'theatres of the War' but who had not mastered the German language. In the Essen section, we had two interpreters and it was an interesting study in techniques of interrogation, to observe and to listen to some of the 'old hands' when they were using intermediaries.

Where language was restricted to simple, calmly delivered phraseology, results were often surpringly successful. As far as I was concerned, I found no great problem in translating back from German to English where necessary, but I avoided like the plague any suggestion of interpreter role-play.

Until, that is to say, on the one occasion that I was prevailed upon by senior colleague, Sgt. Brian Taylor in what he termed, an extreme emergency.

He had an appointment with Heinz Renner the Oberburgermeister of Essen. (The only Communist, I believe to achieve such an exalted role in the British Zone.)

Both Section interpreters were occupied and although I vigorously resisted his request I was finally compelled to undertake the task, albeit with misgivings and only after a long discussion with Brian on the imperative need for use of simplistic speech. Nothing ambiguous! No slang! No English idiom! His promises were unequivocal.

Brian was an embullient character rich in humour and charm and had been very supportive to me when joining the Section. A pipe smoker, who enjoyed a fond, but justifiable perhaps, self image of a 'typical Englishman'.

As we drove into the City centre, I rehearsed some basic phrases of introduction and after numerous reassurances from Brian began to feel more confident. On arrival, we were given the usual degree of respect that German authority accorded us at that time. We were ushered into the Oberburgermeister's office and after the customary handshakes we settled down.

Brian took out pipe and with a measure of theatricality lit up saying (was it thoughtless or malicious?): 'Ask Herr Renner whether he is jumping out of the frying pan into the fire'.

NB. The German version 'out of the rain and into the gutter' did not readily spring to mind on that occasion!

One role in Field Security that proved more satisfying at that time was that of 'Anti-Subversion'.

De-nazification procedures, coupled with expansion of democractic idealism was not easy to deliver to a German nation, for so long subservient to the rigidity and excesses of National Socialism and still reeling from a crushing defeat. There was little doubt that elements remained who wished to retain the twisted nationalism of the previous regime. It was a fact however that no resistance movements of any consequence developed in the immediate post-war era. Thanks, one hopes, to our own vigilance, although this was greatly assisted by the prevailing lassitude of the population.

A short while ago, I noticed a reference in the 'Radio Times' to the '88' Movement. The writer inferred that this organisation had sprung up in Germany in the early fifties as a beginning of an anti-democratic trend. It was an impression which I should perhaps have written to correct. '88's were scrawled ubiquitously on walls and doors and chalked on bomb sites in the early days of the Occupation. The significance, if such it can be termed, originated with the 'Heil Hitler' greeting. The letter 'H' being the eighth in the alphabet – hence '88'. Activity never extended beyond vandalism.

(I sometimes think the naming of Charter 88 in Britain was unfortunate – but possibly it may have served to provide an alternative reference to that above.)

Thoughts of subversive activities do serve to stir memory of an occasion, an unsavoury one, experienced when later I was serving with 415 FSS in Westphalia.

Shortly before this incident, a group of English girls working for the Control Commission had arrived at 9 AIO in Arnsberg and this coincided with an invitation which our Section had received from the German authorities to be their guests at the first post-war appearance of a circus in the town. The entertainment-starved population anticipated a spectacular event.

Our Section, needless to say displayed its customary charm and a complement of the CCG girls acknowledged their pleasure at being asked to share the grand occasion with us.

The layout of the circus ring failed to evoke memories of the pre-war Bertram Mills pageantry. Its radius mesured less than a cricket wicket. A circus in miniature.

The British contingent joined by the local Burgermeister were invited to occupy the close circle of ringside seats as a tribute to our 'eminence'. The Section preened, the CCG purred.

Anticipation grew with the arrival of three white chargers who were coaxed into a leisurely anti-clockwise parade round the tight circumference. Tempo and excitement intensified. At first a trot, then onto a canter and finally to a sweat spraying, tail flying gallop. Faster and faster until the animals were simply hurtling round the arena.

Then it happened. The expression 'hitting the fan' is now commonplace. On that occasion it was a matter of 'hitting the fans!'. Absolute pandemonium reigned. Our speed of evacuation was only equalled by that of the offending white charger who seemed quite unruffled – perhaps 'relieved' would be a fitting adjective.

The long-awaited show disintegrated. We were subsequently showered with apologies (which made a pleasant alternative) but as we retired to the luxury of the 415 accommodation overlooking the River Ruhr we continued to debate whether we had been victims of a brilliantly organised subversive 'movement'.

16 FSS
An Explosive Situation

Bob Maxwell

Towards the end of summer 1947, Major McAllister-Smith, G2 I at Klagenfurt in Austria, tasked 16 FSS, at Wolfsberg in eastern Carinthia, to provide three men, to leave within half an hour, equipped for a week with clothing and personal weapons, to drive to Villach and meet Captain J.M. Dickson at 68 FSS. Under his command they would proceed to Mallnitz at the westem end of the province and take over the investigation into the sabotage of the military leave train from 88 FSS, the local Section. For some reason 88 had blotted their copybook in the immediate aftermath of the incident. I did not know then, nor do I now, what they had done or omitted to do to harvest such displeasure! In addition to myself I detailed Sergeant Rathbone, from the Wolfsberg detachment to pack at once, and phoned Sergeant Souza from the St. Veit a.d. Glan detachment to get ready.

Quarters were arranged for us at the military ski school at Mallnitz, although winter sport or mountaineering were the last things on our minds. The village of Mallnitz was in a virtual cul de sac at the western end of a long valley; it was at the edge of the British Zone of occupation, neighbouring the Americans in Bad Gastein on the western side of the Ankogelgruppe mountains. Beneath the peak of the Gamskarlspitze, was an eight and a half kilometre long railway tunnel linking the two zones.

One side of the valley, not far from the tunnel on the British side, the railway gradient gradually took the trains through Spittal a.d. Drau to Villach and Klagenfurt. Trees, shrubs and bushes on the lower slopes reached upwards virtually to the edge of the rails . This was important for the saboteurs as cover. Although the viaducts look tempting as places where a bomb might best be located, they had no soft soil beneath the tracks to bury any explosives.

The international scene must not be forgotten in this story; sabotage, terrorism and murder were taking place on the distant shores of the Mediterranean as the Stern Gang and the Irgun Zwai Leumi, the latter led by Menachim Begin (later to become a Prime Minister of Israel) committed the outrage of the King David Hotel. Even worse for Intelligence Corps personnel the Irgun had in their fight for the creation of a new Jewish state captured, hanged and booby trapped the bodies of two Corps Sergeants. Just over the mountains in the American Zone there was a camp housing, amongst many others, a large number of Jewish refugees from all over Europe.

One of these had been caught by the Austrian Gendarmerie on the morning following the explosion as he came down from the mountains. He had been armed with an automatic pistol and was being held in a prison under American control for the illegal possession of arms. This capture was not reported to us until some time later as the lines of communication went from Salzburg to the Control Commission in Vienna, back down to Klagenfurt and eventually to Mallnitz.

In the meantime we had been researching in the village itself. Captain Dickson, formerly of the Palestine Police, knew very well the procedure following such incidents with which he was not unfamiliar. We needed troops to seal, if necessary, and to control, certainly, entry and exit from the village. A section of an Irish regiment was assigned to us under a Corporal. Unfortunately the relative freedom from normal military routine, and the alcohol they consumed, went to their heads! One of them posted as telephone orderly had to be physically subdued before being returned to his unit under arrest: the local HQ and Sergeants' Mess was in one of the larger hotels in the village, its telephone system behind the bar within easy reach of a tempting array of bottles – too tempting. At the time none of us was concerned with such a minor detail as human frailty until the aftermath was discovered the next morning . . .

The first four days and nights at Mallnitz were virtually without sleep. We scoured the village for any unusual absentees from their places of work or residence; we interrogated dozens of people, particularly those working in public places like Gasthäuser (pubs), hotels, eating places, bus and railway stations, for anything or anybody out of the ordinary or unusual. Distant farms up steep cart tracks into the mountains were both checked and searched and their occupants questioned about anyone unknown to them passing by. Had their dogs at any time during the recent days or nights barked at strangers or without the farmers necessarily knowing the cause? The dull routine grind of criminal investigation.

Eventually, if we did not strike gold, we got copper! Two young men, not Austrians, had been in one of the Gasthäuser the evening before the explosion. Between them they had been carrying a rucksack packed with something extremely square, like a large tin. According to the landlord they treated it with great caution doing everything possible to keep it out of the way of passing feet and an accidental kick. At this point we brought detectives in from the Gendarmerie in Klagenfurt for their assistance in forensics and identikit pictures of these suspects for circulation. If this rucksack had in fact been carrying the bomb, it was a fair size, measuring approximately 10 ins high by 6 by 8 and probably contained some home made amateur explosive. Anything of that size would need burying beneath the railway line; it most certainly could not have been left just beneath a rail.

These clues seemed to explain the apparent ineptitude of the bombers. Because the bomb had been buried on the outside of the tracks the explosion threw the coaches inwards against the mountainside rather than toppling the whole train down the precipitous slopes into the valley. Additionally the sophistication of modern terrorism and explosives like Semtex were still unknown. The suspects were apparently European refugees, young, and had probably not served in any Army, Allied or Axis.

By this time the capture of an armed man in the American Zone had filtered down to us in Mallnitz. Sergeant Rathbone and I were to go through the tunnel to the US Counter Intelligence Corps in Bad Gastein to discuss the arrangement for his extradiction. They would not have us coming across in Intelligence Corps uniform so we had to remove all badges of rank and regiment and slipped 'Press Correspondent' bands on to our shoulder straps! We were given an extremely cautious reception. They knew, even if we did not then, that the man being held was a Polish Jew. Their international political viewpoint was also in favour of the establishment of a Jewish state and against British imperialism even if the Palestinian Mandate was from the United Nations. The early relationship was therefore a little bit tricky!

Arrangements had been made for a special train to take myself, Sergeant Rathbone, and three or four other volunteers from the ski school staff through the tunnel to collect the suspect. As there was no question that we would be in uniform the train would stop at the small station at the tunnel exit on the American side and we were barred from leaving the station precincts. The Americans would bring the prisoner to us.

After a little wait they arrived by jeep, handed the prisoner over to me against my signature and departed. I gave an order to all present that under no circumstances was the prisoner to be physically abused in any way, that as soon as he was in the coach an all round guard should be kept and that members of the party were to keep out of his personal reach at all times. I then made a mistake by going to the station master to get the train back through the tunnel. When I returned from this errand and entered the coach, all my orders had been obeyed except that the victim was sitting isolated in the centre with blood pouring down his face from a bloody nose. The senior man present was a Staff Sergeant from the ski school who explained, following my livid inquiry, that 'he had tripped on the steps while entering the coach!' Not even to this day will ex-Sergeant Rathbone tell me what really happened.

An interrogation room and cell had already been prepared within Mallnitz station and after ordering the cleaning up of the prisoner at the first tap we could reach, he remained there under intensive interrogation from Sergeants Rathbone and Souza. He had come from the refugee camp outside Bad Gastein in the American Zone, which housed a number of Jewish displaced persons (DPs) among others fortunate enough to have survived the Holocaust. Whether or not he had in fact been one of those who had perpetrated the attempt against the military leave train will never be definitely known.

Eventually the only charge that could be made to stick was illegal possession of a firearm.

But the possibilities of further investigation obviously lay on the American side of the frontier. In Mallnitz itself the operation was over. With the exception of Sergeant Rathbone, we all returned to our respective units. He on the other hand continued the investigation in the US Zone. The American CIC declared that they would neither help nor hinder but gave full access to the Gendarmerie. After many weeks of special secondment to this operation he eventually uncovered cells, records and arms caches in the DP camps but was unsuccessful in the hunt for and conviction of the bombers.

One Man's War (mostly with 19 FSS)

Reginald A. Weaver

The viewpoint of ex-CSM Reginald A. Weaver, 19 FSS was the only Section to land on the Normandy beaches on D Day, 6th June 1944, while the famous 50th (Tyne/Tees) Division which it served, was disbanded in November 1944.

1942

26 July **Liverpool**
SS 'Samaria' and 3000 troops – the pleasure of unexpectely meeting Phil on board the same ship.

31 July **Clyde Gourock**
How wonderful my country looked the night before we sailed: the glorious sunset and how sad I felt to leave all I loved and I knew not for how long.

11 Aug **Freetown**
We were not allowed ashore in this 'the white man's grave'. Here our convoy reformed for the next stage of our voyage.

30 Aug/ 1 Sept **Durban**
Kind people showering hospitality on British troops. How

we spent all our money on food in this land of plenty. The rickshaws – the distinction between whites and blacks. A little Methodist chapel where troops were welcomed for a sing-song and refreshments.

16 Sept **Aden**
Here we coaled – no time for shore leave – very hot.

21 Sept **Port Tewfik**
The end of our voyage and glad to be ashore. From 26 July to 21 September along with 3000 other bodies we lived, or should I say existed on HMS 'Samaria'. Our convoy zigzagged day and night with our escort of destroyers like watchful sheep dogs for ever searching for the lurking U boat – the occasional depth charges – the scare when U boats reported – the cargo ship which developed engine trouble and was left behind with a faithful destroyer. The tropical heat, the unappetising food – 92 of us in one room, all windows blacked out, doors shut and very little ventilation – the horrible stench at night. The hours we spent on the floor playing 'Solo' to while the hours. The ever popular 'Housey – Housey' played by the troops. How we watched the sea, the glorious sunsets and sunrises and dreamt and talked of home.

2 Sept **Helwan**
IC Depot Middle East where we acclimatized ourselves for 9 days. 'Gippo-tummy' from which I suffered and the resulting nights and days spent mostly in latrines. How thin I was and how groggy and depressed I felt.

30 Sept/ **Alexandria**
17 June 43 The Corniche stretching for 13 miles around the beautiful bay of Alexandria. Lovely bathing and cricket team. The little shoe-black on the docks who called me Abou Shanab (father of a moustache) and Ron, Abou Mafish Shanab (father without a moustache). The good friends I made, Khouri, Moufarages, Socrate and Funf. My interesting job in civvies – night raids rounding up escaped Italian PoWs. Ron Taylor a real pal and how sad I felt to say goodbye to so many good friends in Alexandria

when I left to join 19 FSS with the 50th Northumbrian Division.

17/29 June Ataca

Waiting for the word "go". Living in tents in the desert – sand and flies everywhere and in everthing. The peace and quiet of the desert at night broken only by the sound of bagpipes from a Scottish unit nearby.

29 June/ 10 July The Sicily Invasion

On board HMS 'Winchester Castle' bound for Sicily – the Suez Canal – Port Said and then we were off. The Mediterranean filled with shipping – the storm which arose on 9th July and how we all lay in our bunks wondering how we could possibly get ashore in such seas. As if in answer to prayer, the sea suddenly subsided and in the early hours of the morning of the 10th July, the engines stopped and we knew we had arrived. An Italian searchlight suspiciously swept the sea and we held our breath as it crept nearer to our convoy and then, amazingly, was turned out. The tense voice over the intercom calling the troops to man their boats and then the order "away". These LCA (Landing Craft Assault) set off into the darkness with the first wave of lads – I admired these lads. I waded ashore a few hours later with 'Ginny' Anson. The warships were still shelling the coast.

10 July Avola

Where an Italian picked up a 'red devil' (booby trap) meant for one of us, on the town hall steps – he was seriously wounded. Bombing and machine gunning by German planes.

14 July/ 3 Aug Cassabile

The farmyard where I lay whilst a Canadian MO set my dislocated shoulder. The castle where I spent the rest of my stay – the kindness of the Sicilians who looked after me – the figariadas, fresh figs, almonds, old wine. The royal farewell the Sicilians gave to me, 'Alfredo', as I drove out of the castle in a jeep to rejoin the Section.

7 Aug Carlentini

One night of mosquitos.

8 Aug	**Primosola Bridge** Where many of the 50 Div. boys still lie; a famous battle in Sicily Campaign.
8 Aug	**Catania** The first real bomb damage I had seen. The wrecked German planes scattered over the airfield.
12 Aug	**Linguaglossa/Castiglione** Like a huge castle perched on the hill top.
18 Aug	**Francavilla** King for a fortnight. The AMGOT (Allied Military Government of Occupied Territory) officer asked me to take over till his return. I sat in the 'seat of Customs' and as the only British representative there made all the decisions. Sacked the local Facist administration and with the help of a Judge formed a new local non-facist government. Sat in the court with the Judge representing AMGOT. Dined with the Countess. Known as 'Il Tenente' – the Lt.
2/9 Sept	**Ionia** A wonderful billet with balcony overlooking the Mediterranean.
13 Sept	Sailed in American LCT via Palermo to Salerno.
15 Sept	**The Salerno Invasion** Battleships pumping shells into the hills. We moved in at dusk and were only just ashore when the Germans hit and blew up our ship. Dozed near our truck that night – bombed.
17/30 Sept	**Pontecagnano** In the woods with our cardboard deception tanks. Played Bridge day in day out from morning until dusk then under our mosquito nets at night (millions of mosquitos buzzing and bombing).
30 Sept/ 2 Oct	**Salerno to Taranto** A wonderful ride over the mountains – glorious scenery.
8/10 Oct	**Taranto/Messina** A rough sea – a small boat (LCI) – a miserable voyage.

10/17 Oct **Taormina**
A beauty spot high overlooking the Med.

19 Oct **Augusta**
Sailed for Blighty 24 Oct. in HMS 'Otranto'. How happy we felt! Met Ron Wrigley coming aboard.

26/28 Oct **Algiers**
Not allowed to land. Heard of another convoy attacked by U boats near Algiers.

5 Nov **Liverpool**
Joy and impatience. We landed at night and were whisked away to Suffolk. Not for us a welcome back to England.

1944

5 June **The Normandy Invasion**
Again to experience that feeling of uncertainty as to what fate had in store on the morrow. We sailed everyone thinking the same thoughts but none expressing them. We played Bridge to pass the time.

6 June The day ('D' day) – the sea full of shipping – the warships shelling any sign of enemy life. Our LCT (Landing Craft Tanks) was shelled as it nosed its way ashore and we all flattened out on the deck as the shells straddled our ship. My feeling of thanks to the Navy when they silenced the gun. Going ashore at Le Hamel on a 'Rhino' – the sign of recent battle – the dead – the ambulances. The march to our first halt a few miles inland – the noisy night as we lay in a farm shed and listened to the Jerry planes attacking our shipping – the beauty of the tracers (if one could appreciate beauty at such a time) gracefully going up to meet the attacking planes.

7 June Back to the beach to collect our motor bikes.

8/20 June **Bayeux**
6am. It seemed so quiet and yet had only just been captured. Breakfast with Allan Moorhead in the Hotel Lion D'Or. An unpleasant morning stranded with a heavy box full of security documents and the Germans attacking

a few streets away. I hid the box in the cellar and awaited events. Fortunately good old 50 Div. pushed Jerry back. Here we captured several stay-behind agents (spies) left by the Germans. We subsequently received a bouquet from London and our Corps commander. Our OC was promoted as a result.

Bernieres Bocage

The farm building where we slept on the straw along with the farmer's family. The whistle of shells passing to and fro and occasionally stopping very close. The yellow foggy dust everywhere on the roads which was most unpleasant when following tanks.

Villers Bocage

The village which no longer exists – we did not pass through Villers Bocage but over it.

Putanges

Still fighting for the place when we arrived. Memories of crazy FS NCO's nonchalently knocking in our TT FSS direction board to the accompaniment of German spandaus whilst all other troops were taking cover.

Beauvais

Women having their heads shaved on the steps of the town hall.

Sept **Albert**

Liberation days really started here. The town barber was still busy cutting off the hair of all women collaberators as we arrived. Our billet was a German HQ the previous night. Joined in the celebrations at night.

Roubaix

Proudy and I were the first British troops to arrive – the place had just been cleared by the FFI and the dead and wounded were being cleared away. A wonderful reception – people fought to kiss us until we were almost dragged from our bikes. Kisses, wine and autographs (I signed hundreds). My bike was covered in flowers and flags with bottles of champagne pushed into my paniers. More wine then on to

3 Sept **Lille**
More crowds, kisses, autographs and wine. Called at the town hall and then on to

3 Sept **Brussels**
Riding along feeling very happy with the world – my bike gaily decorated and with streamers flying around my neck I arrived in another madly happy place. We stayed at the Hotel Cecil – crowds outside the Hotel and once outside a girl on each arm and then dragged from cafe to cafe, everyone dancing, drinking and singing 'Tipperary'. Nobody seemed to go to bed that night – I returned at 5am.

Malines/Heusden
Moving through to more serious days again.

Bourg Leopold
The big conference and the plan I heard of the future operations. The next big push with airborne troops playing a big part.

Aalst
Here we stopped with the Irish Guards – shells whistled backwards and forwards overhead – the rocket firing Typhoons called up to destroy German tanks holding up advance. We then set off at 6am at speed to break through into

Sept **Eindhoven**
Crowds lined the roads we drove in – the quieter but very sincere welcome given by the Dutch – no kissing but genuine handshakes. We joined up with th 101 US Airborne Div. here. They were feeling isolated having just taken the place and were waiting for assistance – the Yanks gave us a good welcome. The second night we were very heavily bombed dozens of ammunition trucks were hit and shells – exploded during the night. Slept in a ditch that night with the situation decidedly dangerous. Nearly fired on as we came out of the ditch in the morning by the trigger-happy Yanks.

Oct **Oosterhout**

Over the Nijmegen bridge with the notices "Step on it – no speed limit – bridge under shellfire". Nearly took off on my motorbike as I hit the repaired hump in the middle of the bridge. Nights in Oosterhout – candlelight – shelled overnight. Dangerous day trips for a few weeks and then back to Nijmegen for more shelling and bombing until our Div. was pulled out and sent back to

Nov **Dixmude**

A ride of nearly 200 miles in a day on a motorbike in rainy cold winds. My faithful steed was the only bike to make it in a day. A farewell to our many friends in 50 Div. – I woke up the following morning with my pyjamas over my battledress.

Dec **Brussels**

Back again. Expected to have a good Xmas there but the Ardennes offensive started and the Yanks sent a SOS for a few experienced FS sections. Ours went to

Dinant

And four of us moved further on to control the security at the bridge at

Givet

Here I spent a sad, cold snowy Xmas. Refugees crowding the roads coming back. All civilians afraid that the Germans would reach the Meuse. Cold depressing days until Jerry was stopped. The only bright spot was the generous food rations supplied by our US allies. Then we moved on to

Namur

Along with the American CIC we handled civilian security problems.

1945

13 Feb/ 3 Mar **Brussels**

Back again to our Corps depot. Called for again by the Americans so off to

Munchen Gladbach
Heavy bomb damage by our planes. Plenty of work.

Rheydt
Over the new Rhine bridge to

Friedrichsfeld
To 19 US Corps.

Sudkirchen
To join 15 US Cavalry Group and then on the road dashing in and out of places. Usually awakened in the night with "Rolling in half an hour" and then off on motor bikes tucked in behind the American type of Bren Carriers.

Hamborn
17th US Airborne Div. CIC. Lots of work here – three of us on detachment working the town. Here I caught my first Gestapo official. Said goodbye to the CIC ('G' men in uniform).

Bottrop, Werl, Unna
Stayed a week here – four of us attached to a US Inf. Regiment. Arrested over 40 people. Received plenty of bouquets from the Yanks for our work – they were sorry to lose us but we had to rejoin the British 2nd Army to move up to Kiel. A long ride from Unna via Erwitte, Soltau, Osnabrück, Lubeck (we stayed about a week before being allowed to move into Kiel).

Kiel
Arrived Kiel 7th May to stay until demobbed. Here we traced and arrested many Gestapo, SS and SD officials. Before leaving had become CSM 19 FSS.

Demobbed 13 Feb. 1946.

Extracts from the War Diary of 35 FSS including the Salerno landings

(reprinted as originally written)

12 May 43	ZAGHOUAN, Tunisia . . . contacts made with the Controleur Civil, heads of Gendarmerie, municipality, etc. Proclamation issued. HQ of a bn Hermann Goering Regt searched.
13	Arrangements made with Gen Mathnet for collaboration of FS Det with FR civ & mil.
14	Black list compiled with help of authy & informers.
15	Arrest of civs continued & para landings story pieced together: FSO 4th Ind Div (290 Section) called.
16	Investigation made into identity of AC with divebombed & MG colm of PW . . . more section FS NCOs appear, now 7 & dvr.
19	more Arrests made . . . FSO calls on HQ & Gi 56 Div
21	GIII Ib 8 Army calls . . . FSO sees GII Ib Army & new GIII Ib 8 Army at Tunis . . . & GOC 56 Div who agrees to leave Det at Z until inbest. over (para case)
23	Contact with Caid . . . Blacklist amended
26	FSO Sousse calls & takes copy of List & next day 2 of his NCOs att.
31	

1 Jan 43 ZAGHOUAN, Tunisia . . . FSO, 8 FS NCOs, dvr plus 2 NCOs of Sction, SOUSSE.
3 NCOs enquiring into subservice orgs.
Further Arrests made in paratroop enquiry . . . Capts La Jeunesse & Burin & 3 inspectors Arr to assist. On 4th FSO to Tunis & hands in dossier – it is hoped that 'at least 5 of the Blacklist Arrestees will be executed & 15 receive terms of penal servitude.'

5 Liaison with Capt Ripley, FSO 6 Armd Div . . .

7 Hands Z over to Section, SOUSSE.

8 Dep & reaches Gabes.

9 Det returns to sec HQ w HQ 56 Div at Tagiura, nr Gabes. Recce for projected of KG IV.

July 43 Moving back, wef late June?, into Tripolitania . . . on 14th visited 37 Section at Castell Benito. MilSec & trg.
Lcpl Angior mentioned & Cpl Vendeman posted to 274.

7 Aug 43 Moved to tented camp 3km outside Tripoli on Homs road.
Sgt Reed posted to 276 as WOII; Sgt Trent in ex-276; Lcpl Vassalo posted in.
Sgt-Maj & Sgt Whittard at HQ 10 Corps.

Sep 43

1 Tripoli, Libya . . . vehs embarked . . . Lcps Waite & Purse w 169 Bde

2 Sgt Cotter & Cpl Travenen att 2101 Gds Bde HQ & embark w 1 MC

3 FSO (As Watt) Sgt Whittard, Cockell & Trent & Cpl Vassalo embark (WOII Freeman, Cpls Bryant, Oakley to fol.) Cpl Luxton att 167 Bde & Lcpl Angior posted to Helwan (2 NCOs short).

9 0600 Cpl Luxton disembarks Salerno tac HQ 167 Bde
1000 Waite & Purse disembark
1200 Cotter & Travenen disembark
1430 FSO & pty disembark

10 am PW intg, civs rounded up . . . prep to enter Salerno town . . . all dets in &
1500

Sec conc less Luxton & Waite w HQ 56 Div & mov to S
1600
At Div HQ & collect Lcpl Dunster 31 FSS
1730
Enter Salerno town: AMGOT & CC.RR contacted . . . telex disconnected: curfew to be enforced by CC.RR; Post searched & mails sealed:

11 0930
Office/billet estab Via Indipendenzia 27: banks sealed by FSO & GIII Ib . . . FS HQ now GIII Ib, FSO & 5 NCOs 35, 3 NCOs 276 & 3 of 31.
pm
Gds checked & Questura & Municipio searched – meanwhile Det w/Div HQ on Ia tasks & civ controls.

12 Port area recce & labour hiring Arranged with DADL . . . Fascist HQ searched & docs found. Constant Ia work done on div front & passed back. 2 NCOs visit Vietri & order CC.RR to keep civs off roads. Liaison w HQ 94 Area.

13 Docs from Fascio & Questra . . . IT Mil HQ s searched . . . Port Sec, but port being shelled.
2100
Lt Walcott & 4 agents CIC V Army plus 4 NCOs 312 FSS Arr 2 NCOs 35 & one of 31 sent back to Bde HQ, rest on gd & patrol . . . GER threat!

14 Ia info from refugees . . . control post estab at Vietri but CC.RR had deserted under shelling. 1 NCO 31 FSS injured on NC & evac to Div HQ. Cpl Brentini att for docs classification: 1 NCO 31 FSS to Bde.

15 Docs class. begins. Civ controls. X Corps Ia offrs Arr & att HQ . . . FS Capt Coloquhoun, XC LO calls. Liaison w OSS & HQ 56 Div & FSO 46 Div.

16 Refugee controls, Ia intgs: house searches & Arrests.

17 Tps looting in S & not treating CC.RR properly . . . by mortar fire & FS HQ hit & damaged. Lt Walcott dep to Amalfi w 4 CIC & 2 NCOs 312 FSS.

18 Suitable contacts made for Maj MUNTHEA (A Force) . . . Ia info obtained & passed back, mostly to att Ia offrs; house searches; Cpl Bryant Arr: refugees now 100 per day thru HQ FS . . . Capt Rafter & NCO & dvr 276 Arr.

19	WOII & 4 NCOs 276 att . . . docs sorting: Salerno Times distributed to fwd tps. Less mortaring & refugees.
20	House searches . . . vettings, Arrests.
21	Brothel staffs vetted! Control posts estab on all roads out of Salermo. 2 NCOs 31 FSS att. Houses beyond Vietri searched for GER & suspect radio transmitter – NTR.
22	Offices searched . . . PO visited but tps had got in & looted: resealed G2 Ib V Army & G2 CIC called.
23	Various Arrests & house Arrests of Fascists . . . Cpl Travenen att 23 Armd Bde w Lt Walcott & CIC. Plan for search of Cava on entering.
24	Constant liaison w Bde about to capture Cava. All Port Sec info to 276. 15 security prisoners in jail & 241 who live thru frontline! 1600 CAVA entered & ruined bldgs searched.
25	Docs from Cava classified . . . HQ 7 IT Army loc & searched . . . 1000 plus refugees there visited. FSO & GIII Ib collect important CVRA docs in Maiori.
26	Various investigations.
27	Various investigations handed on to 276 FSS. Cpl Waite brought in from 56 Div with Malaria . . . Sgt Cotter replaced him at Div HQ.
28	FS establishes a refugee report centre in conjunc w Town Mayor & AMGOT. Cpl Travenen enters NOCERA & accepts 12 × 75mm guns from CC.RR . . . liaison with Det 7 Armd Div FSS (270 FSS). Col Strangeways MI (6) calls & need for more coord reported to him.
29	Sgt Cotter at Baronissi & obtain Ib info . . . keys to Casa del Fascio to FSO 276. 35 FSS begins handover to 276 of docs & info on area . . . GIII Ib, FSO & 4 NCOs intg suspects all night in Ravello ref radio, but NTR.
30	Handover 35 to 276 of SALERNO completed. Travenen, Purse, Lt Moro & CC.RR to Castellamare to take over from Lt Walcott. WOII Freeman, Cpls Oakley & Knight disembark with tpr.

Oct 43

1	At Salerno, preparing to move on.
2	Arr Sarno . . . interrogating, etc, control of civ mov.
6	Arr Nola . . . ditto.
8	Arr St. Maria CV, set up office.
Dec 43	
1/14	At Teano which was handed over to a Det of 278.
15	To over Sessa Aurunca from 31 Section.
2 Jan 44	Arr Lauro.
4	Capt A.S. Watt posted out to 3 SI Unit.
6	Capt S. Sarluis became FSO
13	Arr Naples.
16	Pozzuoli
20	At sea with invasion force.
21	Arr ANZIO.
Mar 44	
1/9	At ANZIO. Note: Section had been in two parties, one at Anzio & one at Nocera.
10	At sea.
11/16	At Nocera.
17	Deps for Potenza.
18/24	At GraViana (FSO, Sgt Maj Whittard, Bryant, Oakley, Vassalo & Nardini) with Det Altamura (Cope, Luxton, Knight & Rubenstein) & Matera (Travenan, Munro & MM Ianessi)
25	Arr TARANTO
31	Embarked in HMT 'Devonshire'
Apr 44	
3	Arr Port Said & moved into Camp 3, Tahaq w 56 Div HQ.
6	Lcpl E.C. Purse awarded Mentioned in Despatches (for ops in ? ?)
8	Lt-Col Sprunt, Comdt, inspected section.
16	Tom Keen took over from Sam Sarluis . . . Lcpl Vassalo posted to 252.
25	Moved to Quetaa Lines, Mena.
May 44	At MENA . . . GER & IT lessons . . . Cpl Reason posted in.

Jun 44	MENA . . . language return shows that the 11 IC pers, WOII to Lcpl spoke 8 languages to one degree (bilingual to slight).
11 Jul 44	Embarked at ALEX . . . Capt Sam Sarluis rejoined as FSO.
11	Arr Taranto.
22	To Tivoli.
1 Aug 44	Took over PERUGIA from 278 Section: vetting civs in div conc area. Handed Perugia to 310.
14	To Tottentio.
24	Sassoferanto.
31	Urbino but no CI role, as 290 was present.
Sept 44	Moves in Italy with 56 (London) Div . . . dealing mainly with refugees in areas incl Montefiore & Morciano.
Oct 44	Santarangelo . . . in town.
12	Macerata . . . with dets at Pollenza & Tollentino
Nov 44	At Macerata uc 8 Army.
1 Dec 44	At Forlimpopoli, Via Curava.
7	Capt Sarluis dep & Lt R.T. Adams becomes FSO.
17	Entered Faenza & set up Via Severoli 16 . . . later liaison with det of 263; wef 20th with 407 of 10 Ind Div & with 2 NZ Div FSS.
22	Handed over role to 68.
	Returned to div role in fwd bde areas – now restricted by deep snow.
Jan 45	Ditto in Faenza areas; 2 NCOs (Luxton & Oakley to Austria course, FSW).
29	Arr Russi.
Feb 45	At Russi uc 5 Corps.

56 Div area & control of mov across the r. Montone, r. Lamone & Canale di Naviglio . . . wef 27th also franking AMG issued passes for 56 & 8 Ind Div areas. 14 arrests made. Dets at

	Russi & Granarolo; 6 road blocks operated & 27 IT mil uc as gd pers.
	Handed over to Can FSS & 406.
Mar 45	Moving with 56 Div.
11	Contacted 88 Section.
12	Took over FORLI from 88 Section . . . arrests, etc.
	Set up HQ at Forli, Via G. Regnoli with FSO, Sgt-Maj, 1 Sgt & No. 3 Det of 2 NCOs.
to 2 Apr.	
2 Apr 45	Dep Forli, handing over to a Det 47 Section.
3	Arr RAVENNA, HQ at Via G. Mazzini 28, with 3 Det until
9/12	S. Alberto.
13/15	Longastrino.
16	Filo di Agrenta . . . with the 3 dets generally moving also four times in the same period (No. 1 Fiumi Uniti area, Mandriole, Madonna del Bosco to S. Biagio; No 2 only twice to Foraace di Filo & No. 3 from Ravenna with section HQ on 9th to Palazzone with 169 Bde 10/11th; Longastrino 12/14th to Filo di Argenta from 15th.
28	Into ROVIGO, which was handed to 68 Section on 29th.
30	Into VENICE with 'S-targets', a Det at Mestre.
5 May 45	To Cervignano.
6	Gradisca . . . obtaining info on Jugs . . . 2 NCOs posted out: Oakley to 31 & Reason to 88.
16 Jun 45	GRADISCA - Lt E. Nolan became FSO . . . liaison with 96 & 429.
10 Aug 45	WOII Freeman Sgt Whittard demobbed.
27	New WE (II/359/3) permitting more Sgt ranks.
Sept 45	HQ at Monfalcone . . . Muggia Det closed on 17th.
27	Lt D. Huggins became FSO. Sgt Granville posted in, Sgt Luxton demob.
Jan 46	MONFALCONE – Sgt Granville to Sesana Det: Tpr Mitzrahi to transfering.

Feb 46	At MONFALCONE uc Lt Huggins; WOII Travenan, Sgt-Maj. 422 Section dep to Austria.
Mar 46	Cpl R.A. (Rex) Butler & J.C. Dick posted in.
3 May 46	Lt Granville temp FSO.
9	Lt P. McConachie became FSO. General difficulties with 'pro-Slav element': FSO conf at HQ. 13 Corps ref arms dumps, but raids (Op FINGER) by VG Police & FS on 5 locations in M area, but 'only one even partially successful.'
Jun 46	Last war diary entry – NTR. Sgt A.W. Austin posted in ex-427.
From May 45	56 (London) Div had garrisoned the Province of Trieste & 35 Section had been based for most of that time at Monfalcone. Lt Huggins, FSO in early 1946 was a Prof of Leeds Univ & the Sgt-Maj was WOII Travenan, after the wartime FSO & Sgt-Maj had been demobbed. Lt Peter McConochie, formerly Ssgt IC Pola FS Det was the next FSO, followed by Lt E.J. Granville in the last few months, while WOII Rex Butler was the last Sgt-Major. NCOs incl: Sgt J.W. Dickson (went to Austria as an FSO, Sgt Frank Whitfield, Sgt Fred Liebar. (John Granville went as FSO to 429, then came to 35).
	Section disbanded in Italy in 1947.

37 Section. FSW Orders
S Force Operation
Western Desert, Africa – April 1943

Ref. Maps: 1:200,000, Sheets 2, 3, 5 and 6
1:10,000 Town Plan of Tunis (GSGS No. 4272)

Information

1. A Force called 'S' Force has now been formed in order to co-ordinate the requirements of the different intelligence, Consular and other Organisations. Composition of this Force is as shown in Appx No. 1.
2. 'S' Force is an Inter-Service and an Allied Force formed for the purpose of giving assistance to the various organisations interested in the town of TUNIS or BIZERTE.
3. 'S' Force will proceed to Tunis or Bizerte,whichever may fall first at the earliest possible moment after the capture of these towns.
4. Every section will have a representative of the Bureau de Securite Militaire.

Object

1. To obtain the maximum intelligence material and equipment and to facilitate the entry into TUNIS and BIZERTE of those

organisations whose duties necessitate their rapid entry into those towns.

Method

1. Sub-sections will move on 'D' Day with HQ Section at rear of Block No. 4. forming part of the 'S' Force convoy of Tunis.
2. On arriving at the town of TUNIS sub-sections will at once proceed to their tasks as follows:

Sgt. Wilkins 'W' Sub-Section	Sgt. Saunders 'S' Sub-Section	Sgt Gould 'G' Sub-Section

L/Cpls Goddard/Colmans
'X' Sub-Section

On arrival at your target, your instructions are as follows:

1. To see if the building is occupied.
2. If occupied and persons in same are on the Black List, you will:
   (a) Arrest them.
   (b) Search their building in their presence for documents, etc.
   (c) Having searched the building, you will return to section HQ in the Area 074580 with prisoners and documents.
   (d) Should, on arival at your target there be no occupants, you will search the building thoroughly for papers, etc.
   (e) Having completed your target and nothing of immediate importance has come to hand, you will at once report to your second target and proceed to work in the same way, as above.
   (f) Should documents of immediate importance come to hand at your first target, you will proceed to Section HQ before advancing to your second target.

Guards

A certain number of Inf. personnel it is hoped, will be available with each section for guard duties. However, if this is not possible there will be a Bn. at Force HQ, so: should your target pass a building of importance, ie. German or Italian HQ, etc, and you consider a guard necessary, the following operation will take place:

The Senior NCO of Sub-Section will stay with the target and the Junior NCO will be sent to the Section HQ asking for guards. With the request for guards, the following information must be given by Senior NCO from his quick recce of position.

1. Map. Ref of building.
2. No. of floors.
3. No. of entrances to building.
4. If detached or semi-detached, or one of a block.

Force HQ

'S' Force HQ it is hoped will be at the National in the Ave de France, Map. Ref. 073556.

Inter-Communication

Inter-Communication between Force HQ and Section HQ will be made by wireless, therefore wireless sets are now being placed at:

Map Ref. 076556 (X Roads Ave Jules Ferry – Ave de Carthage).

Map Ref. 073570 (X Roads Lafayette – Ave de Paris) and one with the Royal Navy Section of 'S' Force in the docks.

37 Section HQ will be about 1,200 yds. N. of 075570 W/Set. along Ave de Paris map ref.

Map ref codes will be changed as follows:

Eastings		Northings	
01 – 1	16 – 7	52 – 0	58 – 6
05 – 2	17 – 8	53 – 1	59 – 7
06 – 3	18 – 9	54 – 2	60 – 8
07 – 4	19 – x	55 – 3	61 – 9
08 – 5	20 – y	56 – 4	62 – x
09 – 6		57 – 5	

Curfew

Curfew will be ordered by Commanding Officer of 'S' Force from 19.00 hrs. to 06.00 hrs.

Rations

1. Every vehicle will have 7 day's rations per man per vehicle.
2. Every vehicle will have 7 day's water per man per vehicle at the rate of $^{1}/_{2}$-gall. per day.

Petrol

1. Each vehicle will have petrol for 200 miles.

Section Managers

No.	*Address*	*Name*	*Remarks*
1	18 Rue de la Mutuello MUTUELLEVILLE	AGIS HQ and Stadtkommandantur	(3973) (3966)
2.	'BELVEDERE' 34 Rue Lamy	GIS HQ	Ex-American Consul's villa
3.	10 Rue Maillot	AGRIFOFLIO, Pompeo Col.	Italian IAC (4461)
4.	25 Rue Saint Louis, DA	TIXIER VIGNACOURT	French – Avocat. (4860)
6.	5 Passage Berthelon	RICIGNUOLO Biaggio	Italian (4358)
7a	Rue Kleber, DA	MANOLESCUE Mme, DA	Roumanian (4552)
15.	Ave de Paris	Synagogue	W/T (4548)
15a.	123 Ave de Paris	CENTONZE	Chemist (4449)
17.	1 Rue de Courbet	AGRIFOGLIO Col.	Private address (4545)
17a.	15 Rue Besancon	CENTONZE	Chemist (4643)

Appendix No. 1

The remaining personnel other than Field Security Units that make up the 'S' Force are:

1.		HQ, 'S' Force
	RE.	Sub. Sec. 228 Army Fd. Coy.
	Sigs.	Det. 18 Army Group Sigs.
	Medical	Det. RAMC.

Pro.	3 Sec. 200 Pro. Coy. CMP.
RN.	Royal Navy Party.
RAF.	RAF Intelligence Group.
	2788 Squadron RAF Regiment.
FSS.	68 Field Security Section.
	51 Field Security Section.
	62 Field Security Section.
	31 Field Security Section.
	37 Field Security Section.
	55 Field Security Section.
	(Four FS. Sections to come under command at a later date).
BSM.	Bureau de Securite Militaire.
	Two Sections Garde Mobile.
Civil	Civil Tunisian Detachment of Civil Officers and US. Vice Consulates.
PWS.	Psychological Warfare Section.
AFHQ.	Capt. Chamier, Controle Technique, AFHQ.

3. One Inf. Bn. will come under command on arrival 'S' Force at TUNIS.

Time of Signature 14.30 Captain
30th April 1943 O i/c 37 FS. Section.

Distribution:
Capt. EB. Every
CSM Kalis
Sgt. Wilkins
Sgt. Sainlers
Sgt. Gould
War Diary
File.

39 FSS
North West Expeditionary Force
Norway 1940

Desmond le Grand

When I volunteered into the Field Security Corps (which only late became known as the Intelligence Corps) in February 1940, our main training centre was at Mytchett Barracks, Aldershot. An ideal place to footslog, squarebash, pick up a minimum of discipline, learn to salute, fire a .38 pistol and career around the countryside – in or out of convoy – on Norton 500s, BSA 350s and Matchless motorbikes. 'Passing out' and breaking up into sections took place about four weeks later. Some of us moved on to Sheerness where there was soon to be an emergency, namely Norway.

The North West Expeditionary Force (NWEF) was beginning to have a rough time and the Allies were gradually being forced northward by the superiority of German weaponry, in particular their Luftwaffe. There was an urgent need for more field intelligence. We were therefore rapidly mustered into what became known as 39 Section complete with motorbikes, .38 pistols, a few cartridges and the somewhat meagre personal equipment one had to expect in 1940, and bundled off by night train to Edinburgh. My recollection of our two or three days there is a little hazy but one thing I shall never forget is the hot bath which Selwyn Cregeen and I managed to squeeze out of a dumbfounded head receptionist at the North British Hotel in Princes Street ! After this morale-restoring exercise we invested our meagre resources in very large T-bone steaks probably the last we would see for some time.

A day or two later 39 Section embarked on a troopship and made our way in heavy seas to the Norwegian coast, south of the Lofoten Islands, escorted over by a destroyer which, I was told, was HMS Cossack. After an abortive 'pass' at what I believe was Saltfijrd we steamed northward and finally landed at Bödö after creeping along a rather sinister fjord in semi-daylight. At this latitude, at this time of the year, there is no darkness: at midnight one could read quite easily in the leaden shadowless light. After the gruelling crossing in mountainous seas we landed without interference, feeling rather the worse for wear and looking like lemons despite the repeated rum rations on board. Curiously the Germans, who must have been aware of our landing, did not intervene by air or in any other way.

After rapidly settling in and building up contacts with Bödö's few remaining civilian and military representatives, 39 Section got to grips with its most urgent task, ie communications with other 'I' Corps units north and east of Bödö. It was also vital to try to spot German parachutists and supply 'drops' which were regularly being reported by our people and the Norwegians. A number of recces were made round-the-clock during the ensuing days. Useful information was collected but there was no actual contact with the enemy who appeared to be moving up fast from the south. The Luftwaffe was very active in air reconnaissance, and carried out frequent dive-bombing attacks with Stukas on selected targets. Their task was made easier by the lack of darkness and the weakness of allied air defence. The only friendly aircraft we saw during this period were a few RAF Bristol Bulldogs, the occasional Hurricane, and every so often the odd naval aircraft, including the obsolete Fairey Swordfish torpedo carrier. This we invariably greeted with a cheer and a laugh to speed it on its way: speed was the one attribute this otherwise excellent machine lacked! The Bulldogs did their best (two enemy aircraft being shot down near Bödö I believe), but they were hopelessly outnumbered and outpaced by Messerschmitt 109s and 110s whose activities were growing apace. Bödö's small harbour installations – used essentially by the local fishing fleet – attracted special attention and the jetties suffered heavily. Bödö's main activity in those days was fish salting and drying, and after each raid the pungent smell of salted fish, which in this small town was all-pervading, became even more pronounced, and is something I shall never forget.

A few days after landing I was allotted my one and only NWEF mission. This consisted of getting an urgent message (which presumably could not be passed by radio) by motorbike to an Allied 'receiving point' well east of Bödö. To get there I had to negotiate fairly hilly and snow-covered country. Unfortunately I was given, as a guide (!) a mileage well beyond the correct one. 'Getting nowhere fast' I remember pulling up at an isolated farm to try to sort things out. As I gingerly entered the barn I spotted a large BMW motorbike inside. In a flash I imagined a German paratrooper getting his sights on me from the shrubbery so I sprinted back to my steed and made off fast but no wiser. My problem was to identify my destination in what seemed to me to be hostile country. I tried several itineraries, one of which provoked the sudden appearance of what I took to be a Swedish flag. Some shots were fired, and to this day I do not know if the flag was genuine or whether the shots came from a Norwegian unit or from a German paracommando in disguise. I doubled back on my tracks and finally reached my destination.

The return trip to Bödö was less lucky. I decided to take a mountain pass which appeared to me to shorten the mileage but this road had been bombed (by mistake?) by Stukas which were obviously tryingto hit a village in the valley below. Unsuspectingly, I careered round a bend and was suddenly confronted by large boulders and craters. The crash must have been severe because I remember nothing of it: I was picked up an hour or so later by one of our patrols and taken to Bödö hospital with multiple fractures in my right leg, a torn left shoulder and severe concussion.

The Norwegian doctors and nurses were remarkably efficient and fixed me up as best they could with stirrups, wires and weights in a specially fitted bed but my stay in Bödö hospital was far from peaceful. The German hold on Narvik and surrounding villages was weakening as a result of Royal Navy and Allied ground pressure, and the Luftwaffe was becoming increasingly 'bomb happy' over Northern Norway. We were treated to Stuka/Messerschmitt visits almost every day. As Bödö was built mainly on rocky foundations the effect at ground level of these raids was very destructive . In my ward of eight or nine beds, we used to apply 'bedpan drill' as soon as an airraidwarning sounded. Bedpan drill consisted of placing the utensil over one's head as a protection against flying glass and other debris. If, as happened occasionally, a bedpan had been in use before the warning, ironical messages of sympathy were bellowed across the ward to the unfortunate owner. However our nurses anticipated this problem remarkably well! During some of these raids one could hear the occasional 'crack' of a shotgun outside the hospital precincts. As I was flat on my back I could see nothing, but was told that the shots came from the hospital gardener who was venting his feelings with his 16 bore . . .

After a few days the raids became fiercer. Allied pressure on Narvik was reaching a climax and it seems the Germans suspected the British High Command of operating a communications centre from the basement of the hospital. This was quite untrue but the accusation provided a good excuse for the Luftwaffe to have a major swipe at Bödö on 27 May. The attack was vicious and indiscriminate and was carried out by a combined force of Stukas and Messerschmitt fighter-bombers. The hospital was hit repeatedly by HE and incendiary bombs, and was soon on fire. Norwegian doctors and nursing staff, and our own people, who were trying to rescue casualties outside the building, were strafed by machinegun fire repeatedly. I was told later that the hospital staff suffered severe losses. When it became obvious that the building would have to be evacuated quickly, all my ward mates were moved to the ground floor. As I was tied to my bed by stirrup fixtures and weights, I could not follow them, so I asked to be moved away from the large window at my rear! I also requested a bedpan and all the pillows available, and proceeded to cogitate on my probable fate. Extinction by explosive – or the frying pan?

After what seemed a very long time, an RAF Flight Lieutenant Store (a South African), and a young NCO whose name I cannot recall, reappeared with two or three other chaps and a stretcher. How Store – who had shrapnel in his foot and the NCO – who had been hit in the throat – managed the feat I do not know, but my debt of gratitude to them is immense. Meanwhile the building was being rocked by explosions and was burning fiercely. There was plaster dust everywhere and we all looked like ghosts.

Since it was no longer possible to remain in the building, the bedridden casualties like myself were moved outside and placed as near to the walls as possible. Removing me from my ward bed in a hurry had made it necessary to disconnect my leg extensions and the ironmongery that went with them. As a result my leg began to swell enormously – and shorten by inches – due to unsupported fractures. However this was a relatively minor problem compared with our exposure to Luftwaffe strafing. But the attacks gradually abated and it was decided to move the stretcher cases away from the centre of the town which, by now, was completely ablaze. Since most of the houses were log or timber-built it is easy to imagine the population's ordeal and the extent of the disaster. A few snaps taken by a valid member of our Section (Selwyn Cregeen I believe) illustrate perfectly these dramatic moments. Another casualty and myself were finally heaved into an old taxi and driven through cratered streets to the outskirts of Bödö to get away from the inferno. The driver decided to 'park' us in a nearby log chalet inhabited by a young couple. I was laid down in the hall and someone deposited a large carton of Woodbines on my chest presumably to cheer me up. Looking down the corridor I could see several other earlier arrivals in a much worse condition than myself. One poor devil had been placed on a divan in the lounge and was bleeding profusely from hip injuries. Our young hosts did their very best to assist all and sundry; they were quite splendid despite the havoc that had suddenly hit them. As the raid abated, a Norwegian doctor suggested that we should be moved to a nearby more or less intact psychiatric hospital where there was more room and better medical facilities. So we were moved there in what remained of Bödö's public transport. On arrival – most of us still covered with plaster dust and looking very spooky indeed – we were greeted by the inmates with a concert of grunts and howls: their behaviour had, of course, not been improved by the raid.

After a perfunctory clean-up and essential first aid we were loaded, around midnight, onto an old lorry (which, judging by the shake-up we received, must have been running on flat tyres) and taken to what was left of Bödö dockland. We were then hurriedly transferred (the Germans were moving up fast) onto a frigate whose destination turned out to be Harstad, north of Narvik. Apart from a few air attacks and U-boat warnings, the run up the coast was uneventful. In Harstad the Queen Alexandra Nursing Service – which had arrived a few days earlier – took over. Our canvas field hospital had been set up outside Harstad on high ground to avoid raids as far as possible. We were not the first arrivals. The place sheltered a miscellaneous collection of nationalities, mainly British, French 'Chasseurs alpins', Foreign Legionnaires, Norwegians and Poles. Amusingly, there was even a German Heinkel pilot who had been shot down over Harstad by, I was told, a Bristol Bulldog. He kept on claiming that he had not been shot down but had crashed due to engine trouble. This drew jeers from all quarters!

Some of the casualties were in a pretty bad way. A French 'Chasseur alpin' in the bed opposite mine died from chest injuries. A nearby Polish volunteer had been shot through the forehead and seemed to be out of his mind as he talked incessantly. A young Scot in the bed next to mine had been hit repeatedly in the thigh and was facing amputation. When my turn came to go to the operating theatre an oversight occurred in that I was given a large meal just before O hour. Inevitably, the anaesthetics brought up my lunch, much to the disgust of the doctors. X-rays were not possible because although the equipment had arrived safely on one ship, the X-ray plates were on another which had gone down. The orthopaedic surgeon (a Harley Street man) who dealt with my leg, worked wonders by hand manipulations with RAMC assistants, and 'got things straight' literally. I was then put in plaster. Later, in Scotland, the doctors marvelled at this feat of blind manipulation. So have I to this day.

Our stay in Harstad (two to three days) came to an abrupt end when we were told that we were returning to England in hospital ship SS ABA. I will not hide the fact that the news was welcome! Reports from the western front were becoming increasingly depressing, and although we had all but surrounded Narvik, the situation in the south was indeed bleak.

Unexpected complications with my leg prevented me from returning to the I Corps and ultimately joining an OCTU. I was finally invalided out in June 1941. And so ended my short and somewhat ill-fated contribution to the 1940 Norwegian campaign.

50 Field Security Section 1942–45

John Green

The War Service of 50 FSS can be easily but unusually summed up in three words: "London – Brussels – Berlin". In spite of this, it was always an active 'field' section, only bound to an office by the normal demands of administration and direction.

January 1942 saw the Section attached to GHQ Home Forces at St Paul's School in Hammersmith, under the command of Captain Jack Beausire, ex-Unilever. It was responsible for the overall security of three buildings, the most important being Storey's Gate in Whitehall, used by the majority of GHQ staff and the War Cabinet. One NCO was always on duty near the King Charles Street entrance from 8am to 8pm, supervising Treasury doorkeepers and dealing with any problems or crises. Besides a duty NCO at St Paul's School, another was on detachment for two weeks at a time at Wentworth Golf Clubhouse which had underneath it a complete range of offices and equipment available as an emergency GHQ in the event of widespread destination in London.

Over and above these ongoing responsibilities, NCOs were often called upon by GHQ and MI5 for special duties in London and elsewhere in the Home Counties. These included surveillance of individuals or crowds, particularly political meetings in places like Trafalgar Square. Another task was to oversee the security of the trials of the DD (swimming) tanks on Camber Sands and inland lakes; this usually entailed escorting them from and back to the factory, as well as sharing day and night watches with military police.

A request from M15 resulted in seven NCOs being sent on an SOE sabotage course at their Special Training School 17, slanted towards Public Utilities. This operation, under the direction of Lord Rothschild, was aimed at testing over a period of 18 months the major public utilities (mainly power stations) in the British Isles. Conditions were as near as possible to those any enemy saboteur would have faced. The NCOs, working in pairs, were only given the name of the target and, if present, the existence of a lethal fence. They were instructed to test it by day as well as by night, and allowed a week for each station, which received absolutely no warning. They usually faced 9ft 'unclimbable' fencing (often electrified), dogs, trigger-happy Home Guards and endless locked doors to turbines, pumps and other vital equipment. It required agility, fitness, ingenuity, a lot of bluff and a dose of courage. Dummy bombs were left on specific targets. Such was the high standard of training that the saboteurs seldom failed. If caught, they were instructed to escape or hold out as long as possible. This usually meant until they were thrown into a police cell and interrogated by some senior CID officer when a 'phone call to the area RSLO eventually brought release.

By February 1944 General Montgomery had arrived and GHQ Home Forces had become HQ 21 Army Group. The tempo quickened and culminated in the two-day 'Overlord' Conference in April, with a complete ground floor plan of the Normandy invasion area, for which 50 FSS was responsible in the matter of day to day security. The King, Winston Churchill and other members of the War Cabinet were among the select few who attended. NCOs were permanently on duty at the door and two slept each night in the lecture room with the invasion plan. Captain Michael Pertwee, current FSO and well-known playwright, was present during the expositions and collected as mementos two cigar butts left by the Prime Minister, auctioned years later by Sotheby's for a considerable sum!

May and June saw most of 50 FSS attached to Advanced HQ 21 Army Group at Southwick House, Portsmouth, carrying on its usual security role at HQ as well as a more general one in that sensitive area, crowded with troops. At that point, one of our best linguists, Sgt Norman Kirby, left on secondment to Monty's Tac HQ to be responsible for the security of the Army Group Commander and his tiny operational HQ. This he did with distinction until the German surrender at Luneburg. Other tasks included the return of 30 angry French farmers to the American sector and escorting suspected enemy agents or collaborators across the Channel by MTB or Avro Anson. Early August found the Section billetted in the stables of a Normandy chateau but soon to be detached on September 2nd to join with all possible speed General Horrocks' 30 Corps on the outskirts of Brussels. Having reported to General Horrocks at the Royal Palace, Laeken, on September 4th, the Section joined R Force (a mixed force of specialists) at the Grand Hotel, Boulevard Anspach, and so began its finest hour in a city crazy with excitement at its liberation 24 hours earlier.

50 FSS was designated as the main operational Field Security Section in the city, working under the direction of the Civil Security Liaison Officer (Belgium), Major Donald Loudoun, whose offices were a few doors away. To this end, the Section was quickly allocated No 128 Avenue Louise, a six-storey block of modern flats previously occupied by the Flemish branch of the Gestapo and containing some convenient cells on the top floor. It was a lovely and grand avenue near the city centre and the Section had the good fortune to occupy it for nearly a year till it departed for Berlin.

Its immediate tasks were twofold: (a) to follow up names and addresses on the Black Lists, whilst theoretically making use of White Lists (reliable informants); and (b) to receive hot information. These tasks should be seen in the context of Brussels having been the HQ for the German Intelligence Services in The Low Countries. Both were far from easy in a city that had taken to the streets and was wild with excitement. The Gendarmerie and the civil police had yet to appear in any organised form and small bands of various Resistance Groups were roaming the city, seeking out collaborators and often claiming local military successes against the Germans before they departed. In addition every Bruxellois wished immediately to entertain a British soldier. In view of this, we all had the impression that, for the first four weeks, we worked day and night. It was the experience of a lifetime.

From the earliest day an Information Room was opened on the ground floor of 128 Avenue Louise and people and information poured in. It was open daily from 8am to 8pm and manned continuously by four NCOs (usually linguists). 100–120 persons passed through each day for the first six weeks. Callers and information varied immensely, but all had to be processed. Many individuals were re-directed elsewhere, perhaps to the Town Major, to the civil or military police or just sent home. One Irish lady called regularly with a cherry tart (her "*specialité maison*") for Winston Churchill. Others did have important intelligence and yet others were arrested and sent upstairs for further interrogation, often by the Special Interrogation Unit on the top floor.

Occasionally an individual was brought in under arrest or suspicion more than once. One such person was Winnie Saunders, a large blonde aged 30, who had arrived in Brussels with a US officer in a jeep, having been allowed her freedom in Cologne and Luxembourg all the war in spite of being an English subject. She could not explain why, when interrogated. Having been released and given a job at the Officers' Club, she was again referred to 50 FSS by the Club and again released after further interrogation and was last heard of married to an Australian journalist. Maybe she knew how to play her cards! Some individuals became self-appointed regular informants. Such a one was Victor Slaes whose card declared him to be porter at two night clubs in the city; he was small, sallow and talked huskily in a mixture of English, French and Flemish, always exuding an overwhelming aroma of garlic and being the bearer of *hot* news.

However, the majority of NCOs were heavily involved in an enormous variety of Security tasks in the city, almost entirely with civilians. Using their motor-cycles and two commandeered civilian cars, as well as the Section jeep, NCOs carried out endless searches and investigations, involving WTs, morse noises, lights at night, suspicious uniforms and individuals, vetting personnel at the Town Major's office and incidents such as a time-bomb at the Plaza Hotel (used for service personnel on leave). At the end of six months, the NCOs knew extremely well the streets and places of Brussels as well as the restaurants and the welcoming, hospitable residents. Christmas 1944 was memorable for the Ardennes Offensive (which caused some panic among the Bruxellois) and the possibility of German parachutists landing in the area to link up with the panzers. The code word '*milk*' came through in the middle of the Section's Christmas party and there was a general rush to defend the building with any arms available, only to find later that it was a false alarm.

Of the enemy agents arrested, perhaps the best known and most exciting was Pierre Wilhelm, a Belgian by birth and residence (codename 'George'). He had a reputation with British Intelligence as a leader and organiser of resistance/guerrilla groups and had been expected for some months before he finally surfaced in March 1945. Numerous midnight vigils in remote spots round Brussels had produced no one. Finally a report came into the Screen of a man in a city bar claiming to be a parachutist and wishing to contact British Intelligence. He was accordingly visited in his hotel and proceeded to relate his story to a surprised officer and NCO, but not before calling for a bottle of red wine to share with them. Educated, most capable and likeable and a veteran of two earlier missions, George had dropped four days previously outside Brussels but had already decided to see his family and then give himself up. The advance of the Allies had undermined his training and dedication.

Inevitably the Section also acquired a select few Bruxellois of proven ability and status who unofficially gave much valuable advice and guidance. Outstanding among them was Madame Anne Brusselmans MBE, a housewife with two young children under ten. With an English mother, a French father and a Belgian husband, she was the only leader of the Ligne Comete (Escape Route) in Brussels not to have been arrested by the Gestapo by September 3rd 1944. For exactly four years, Anne had been responsible for the evasion from capture of over 600 airmen, of whom 176 had stayed in her flat. Only courage, determination and resourcefulness of the highest order had enabled her to survive. Not surprisingly she was awarded a high American decoration as well as the MBE, and was the subject of Eamonn Andrews' 'This is Your Life' and no less than two books. The Section were proud to have had her as friend and adviser during their stay in the city.

This stay drew to a close in August 1945. Half the original personnel had by this time departed in various directions. At the beginning, the Section was able to offer a fair linguistic range, including French, Flemish, German, Dutch, Spanish and a little Polish.

Two members, CSM Stockdale and myself, had been given direct commissions in the field by Christmas 1944 and two more, Sgts Denys Clarke and Ken Self, had been commissioned by August 1945, all for "outstanding Intelligence work" in the city. FSO Michael Pertwee was posted to Cologne as an Area Security Officer. Depleted, but not downhearted, 50 FSS became an FSRD and were posted to Berlin to join other Field Security units in August 1945. Their reputation as "a capital Section" in both senses of the phrase continued to the end.

60 FSS
Germany – Capitulation and Occupation

W.C. Young
O/C 60 FSS, April 1945 – May 1946

I joined 60 FSS in March 1945, after six years in the Royal Artillery, having been commissioned in Dec. 1939 from Cambridge University OTC, where I had spent a year reading Modern Languages (French and German) at Clare College. My previous service had been as a Troop Officer in 143 Fd Regiment RA (West Kent Yeomanry), Command Post Officer (Lt), and then 2 years as Intelligence Officer, Royal Artillery, at the Divisional HQ of 49 (WR) Div. (Capt). In January 1945 I had asked for transfer to the Intelligence Corps on the basis of my languages, and after a three-week course at Matlock, had been posted to the FSS of my old Division.

The Section had wintered in Nijmegen and Capt. Richard (Dick) Williams was FSO. The Section role was largely dealing with the numerous Dutch and other nationality refugees from the north (German occupied) part of Holland. One exciting occasion I remember was when we had been warned that a class of Dutch SS were due shortly to pass out of their Training School in German occupied territory and around ten of them were being sent on a sabotage mission to the South, disguised as civilians and with the objective of a rendezvous with a submarine in the Scheld estuary which would bring the sabotage supplies.

Sgt-Major John Davidson was on duty in the office when two male civilians were brought in. He interrogated the first and was satisfied with his story. However, on calling the second to his office, this individual had decided to come clean and confessed that he was one of the sabotage group and that his companion was the person in charge. This one firmly resisted any idea that this was his mission and it was only when we searched his kit and found in the handle of his razor the standard German equipment of a poison phial to be taken in an emergency, that we realised we had got the right man. They were sent back to Corps under very close and careful arrest.

Dick Williams was promoted Area Security Officer as a Major in April 1945 and I took over the Section just before the Spring and final assault on Germany began with the crossing of the Rhine at Wesel, and the rapid deployment of the armoured divisions across the North German plain to reach the Baltic before the Russians got there, thus heralding the end of the war in May 1945.

49 (WR) Div., still under command of 1 Canadian Corps, were given the task of advancing into and clearing Holland with the 1st Canadian Division. 60 F SS, with some Dutch interpreters, crossed the pontoon Bailey Bridge at Wesel and proceeded westwards through a much-damaged Arnhem, to near Ede just a few miles to the west.

At this stage we were ordered to halt our advance, much to our annoyance, as we had the Germans on the run. The situation in Holland for the Germans was very bad and food for the Dutch civilians was extremely short, some existing on as little as 500 calories a day from whatever they could find. The Germans had threatened to flood the low-lying country and the Dutch Government, in exile in London, asked the Allies to halt temporarily while negotiations for the surrender of the German forces were begun.

At the schoolroom in Ede, the first German staff officers came over under a white flag and discussed the position with representatives from 21 Army Group. 60 FSS were responsible for security at the school and patrolled the buildings and playground.

A few days later, Maj. Gen. de Guingand, Chief of Staff of 21 Army Group, accompanied by Prince Bernhardt of the Netherlands, came to the school in Wageningen on the Rhine, while Capt Young of 60

FSS was sent out into the no-man's land between the front posts to bring in the German party to sign the surrender, induding Seyss-Inquart, who had been the overall German Commander in Holland, and General Blaskowitz, the German Army Commander. It was an uncanny experience pacing up and down the school corridor with a German Staff Captain while the negotiators discussed and signed the document inside.

Following this, much needed supplies of food were dropped by air at arranged points over occupied north Holland, while regular convoys of lorries brought up Army rations which were trans-shipped to vehicles given to the Dutch each day.

The occupation of western Holland followed, for the German forces to surrender, 49 Div. moved first into the eastern area, to be followed by 1st Canadian Division who proceeded to Rotterdam, The Hague, and Amsterdam on the coast.

60 FSS were ordered to proceed and install themselves in Hilversum. The Section was divided into several groups, each with a Dutch interpreter, and allocated a specific task. These included contacting the Dutch town Administration and the Dutch Underground, endeavouring to capture the German and Dutch SS and Gestapo units, and (most important) finding comfortable accommodation.

It was not until we had left the convoys on the main west road and turned off to Hilversum that we realised that we were the first British liberating troops along this route and into the town of Hilversum. We all received an overwhelmingly emotional reception from the Dutch civilians who crowded the streets and gave everyone a rapturous welcome.

The Dutch Underground sent a liaison officer, Wilhelm de Beaufort, who took Captain Young and his party to their secret HQ in a private house in the residential district of Laren where there was a special Guard of Honour to be inspected beneath the Dutch national flag. Plans were made for road blocks to be set up and manned by the Underground members to prevent any civilian movement for some days while the situation was assessed and the extreme food shortage dealt with by supplies being rapidly brought in from the south.

The group targeted on the Gestapo arrived outside their HQ, known from information gleaned by the Section in Nijmegen, to find the members of the unit in the process of walking down to their cars outside loaded with loot and currency, and so were able to arrest them all, luckily without a fight.

Since we knew that Hilversum was to be the Corps HQ and that they would almost certainly requisition the main hotel, the Gooiland, much used by radio stars working at the Hilversum Radio Station (it became an officers' club), the accommodation party was sent to take over the second hotel, the Hof van Holland, owned and run by a Dutch family. Each person in the Section had a private room with wash basin; the hotel staff cooked all our meals from Army rations in return for being fed by us, and everyone lived in supreme luxury after having roughed it for many months since D-Day.

The Section spent a month there, cooperating with the Underground and looking particularly for German or Dutch SS or Gestapo who may have been in hiding or endeavouring to slip back to Germany. All German troops who had surrendered were disarmed, searched for loot and then marched back to Germany. It was a very pleasant month of celebrating with the Dutch and making many friendships, some of which still exist at the time of writing, 49 years later.

At the end of May, after nearly a month in Holland, the Section moved to Germany, taking over from an American Section the Ruhr industrial city of Dortmund, and the towns north and south of it. Practically all of Dortmund, except for one up-market residential area to the south where we had our quarters and office, had been destroyed in the Allied bombing raids either by high explosives or fire bombs. The streets and houses, nearly all gutted, were filled with rubble and the inhabitants were crowded into basements. The Section occupied two undamaged and furnished houses: one for sleeping quarters, and the other for offices, feeding, and for the CO's accommodation. We immediately requested German staff from the Arbeitsamt and were lucky enough to get the chef of one of the best hotels in the city (out of action by bombing) together with his wife and daughter, to do the cooking and housework. It was remarkable what they did with the Army rations and other food brought in by ourselves. For them, although being paid by the Military Government, the main attraction was being fed, as supplies to German civilians were very short.

Section members worked in pairs with responsibility for their areas (i.e.Bochum, Dortmund and Witten), getting to know the local police, administrators, and businessmen, and endeavouring to discover any Nazis, particularly SS, returning to their homes, although not many were apprehended, the more guilty ones having got out of the country or gone into hiding. We did, however, arrest a number of Dortmund Steel Magnates who had worked with the Nazi regime.

As Military Government rebuilt the town and the City Administration, we would vet their nominees to ensure they had no malevolent Nazi background: most had belonged to the Nazi Party to safeguard their jobs, and it was not surprising that very few ever admitted to having supported the National Socialist Party. Later, towards the end of 1945, our attention was directed more to monitoring the activities of the newly-formed political parties and trade unions, and to reporting on their activities.

Demobilisation plans were starting, and each individual received his demobilisation date based on the length of his service, i.e. first in, first out. For instance, the CO, W.C. Young, who had been commissioned in December 1939, was released in May 1946. As a result, turnover of the Section from the beginning of 1946 was rapid, the numbers were running down quite rapidly and the Section was finally disbanded towards the end of that year.

Consistently, the personnel of the Section were all of high IQ, and well educated, many with university degrees or coming from responsible positions in government, local government or business. Most were officer material: some had been offered commissions and had refused, most had not seriously been considered or recommended. Consequently, with such well educated and intelligent people, commanding such a Section was a stimulating experience, demanding positive leadership and firm but sensible discipline, and all extremely rewarding and good fun.

60 FSS
The Normandy Invasion – Before and After

Bill Young, Jack Mason, Ken Ireland, Alan Jordan, John Marshall and David Rubin

The 49th Infantry Division, including 60 Field Security Section, returned to the UK from Iceland in early summer 1942. It was based in Herefordshire and the Welsh Border where it began the long period of training leading up to the Normandy Invasion.

60 FSS HQ was in Hereford itself and the Section, which during its time in Iceland had laid emphasis on Scandinavian speakers, was gradually reformed with French and German speakers newly trained at Winchester and Matlock. The CO at this time was Capt Leslie. Most members of the Section were attached to one of the three Brigades which made up 49 Div., i.e. 146, 147 and 70 Brigades, and reported weekly to Section HQ on their work which consisted of talks on Security to units of the Brigades, the investigation of incidents reported by the units and of instances of careless talk. The Section also took part in Divisional exercises of which the biggest was 'Spartan' held in the Spring of 1943.

From Herefordshire, 49 Div. moved in the Autumn of 1943 to Scotland for special landing-craft and invasion training at Inveraray and Rothsay. (A similar move earlier in the year had proved to be premature.) 60 FSS HQ was in Troon but most personnel were again attached to their Brigades which were spread over southern Scotland. Security courses were held in Largs and Edinburgh. Capt Burgess took over as CO in Troon but remained with the Section for only a few months. 49 Division was, in fact, being trained for the N. African landings but a political decision was made to replace it with a Canadian Division. Capt. Dick Williams took over as CO and went with the Section into France.

Early in 1944, 49 Div. moved down to Norfolk for its final period of training before the Invasion and 60 FSS had its HQ in Norwich. In June 1944, 49 Div. embarked from various ports in southern England and landed in Normandy shortly after D-Day. With the 15th Scottish Division it made up the British Army 30 Corps, taking over from the 50th (Northumbrian) Division which had made the initial landing on D-Day. 60 FSS was put ashore at Courseulles, a small village between Bayeux and Caen and established its HQ in the Allied Bridgehead at Ducy St. Marguerite. Here CSM Grimsey was recalled to England and was succeeded by John Davidson. The Section remained within the Bridgehead for 6 weeks while military supplies were built up, and during this time it made contact with French informants, dealt with security incidents reported by the units, and manned check-points for vetting civilians displaced by the fighting which flushed out a number of 'Line-Crossers' seeking information about the disposition of allied units. At this time, and throughout the campaign in Europe, many of the incidents the Section was called on to investigate proved to involve collaborators who were not of security interest. Their investigations were, nevertheless, often time consuming.

49 Division was involved in heavy fighting round Tilly-sur-Seulles but after the unsuccessful attack by three Armoured Divisions east of Caen, 49 Division was moved over to the left flank under the command of the 1st Canadian Corps. Finally, the Allies broke out of the Bridgehead and almost succeeded in surrounding the retreating Germans at the Falaise Gap. Then began a methodical advance to the Seine, while the Armoured forces made a hectic dash across the north of France into Belgium and Holland with the Germans in full retreat. On the way to the Seine, the Section set up more check-points and arrested several line-crossers and an agent equipped with a radio-transmitter at Mezidon. At this time the Allied Command was hopeful

of crossing the Rhine and ending the war before the winter. 49 Div., however, was diverted to invest Le Havre where the Germans were still holding out. While the Section was at Pont Audemer, south of the Seine, Sgts Ireland and Luxemburg crossed the river with the help of members of the French Resistance and obtained information about the Le Havre gun position for 1st Corps. They also picked up another spy equipped with a radio-transmitter. 60 FSS was next based in Bolbec for two weeks and engaged in vetting the stream of civilian refugees from the area. Here, as throughout the campaign, they were helped by the 'Black List' of known spies, saboteurs, etc. compiled through the Allies' own Intelligence services.

After the fall of Le Havre, 49 Div. again set off towards the Rhine. The Section passed through Valenciennes and Brussels where the Belgian capital was still celebrating its liberation, and then continued through Belgium, with halts at Lierre and Turnhout and into south Holland. 49 Div. was again involved in heavy fighting at Roosendaal after which the Section proceeded to Nijmegen which it reached early in November. By this time the hopes of crossing the Rhine that Autumn had had to be abandoned because of operation "Market Garden"'s failure to capture the Arnhem Bridge. 60 FSS and other units of 49 Div. were thus destined to remain in Nijmegen until the following Spring and became referred as the Nijmegen Home Guard. Nevertheless, this period was not without incident.

The bridge over the Rhine (called the Waal in Holland), and the town of Nijmegen itself were within range of German artillery and Sgt. Ken Ireland was wounded by shrapnel on the steps of Section HQ. The 10-mile strip of land between Nijmegen on the Waal and Arnhem on the Lek was no-man's land and 60 FSS personnel were required to cross the Waal Bridge daily to visit an outpost and check on possible infiltration by the Germans. The north of Holland was still in German hands and the Section was called upon to vet a continuous stream of refugees from the occupied area. Since Sgt. John Marshall was the Section's only Dutch speaker a number of Dutch soldiers were recruited as interpreters. They proved of great assistance and some of them later accompanied the Section into Germany.

The German counter-attack at Bastogne at Christmas 1944 which resulted in the 'Battle of the Bulge', kept all the Allied forces on the west bank of the Rhine on continuous alert, and vigilance was rewarded when an attempt by a team of saboteurs to sail a small boat down the Rhine and blow up the Waal Bridge was foiled by the Dutch Resistance and 60 FSS handled the interrogation of the saboteurs. Nijmegen housed the Headquarters of the Dutch National Police in the Annastichting which, during the German occupation had been directly controlled by the Germans. Perusal of records in the Annastichting by Sgt Jack Mason of 60 FSS and a Canadian Security Sgt., extracted a list of 400 names of Sicherheitsdienst, Gestapo and Abwehr staff and their locations in Holland. The list was circulated by Corps and when the north of Holland was eventually liberated, 60 FSS itself had the pleasure of arresting the Gestapo unit in Hilversum whose HQ the Section took over for its own use.

The Spring of 1945 brought an end to the Section's static role. Capt. Dick Williams was awarded the Orange Nassau medal by the Dutch Government and, on his promotion to the rank of Major, left the Section which was taken over by Capt. Bill Young. CSM John Davidson was awarded a commission in the field and also left the Section; Jack Mason took over as CSM. The Allies crossed the Rhine and 60 FSS moved into north Holland.

Three Incidents whilst with 62 FSS and Trieste Security Office

R.M. 'Dickie' Richards

Banker wins

I was visited in my office one day by a hauntingly lovely woman who introduced herself as the wife of Karl ————. Karl had been a Gestapo Kriminaloberinspektor and therefore, at the war's end, automatically arrested and interned pending investigation as to whether or not he had commited war crimes. Karl had recently escaped from his American internment camp in Germany (which I knew) and had managed to get into Italy where he was in hiding (which I did not know). She would not say where in Italy, but Karl had got a message to her asking her to join him, naming the contacts who would smuggle her over the frontier. She wanted to go, but only to persuade him to come back and face the music, as they both knew he was innocent of war crimes and she wanted to rebuild their lives in their native Austria. Would I provide her with the documents to go to Italy legally and persuade him to come back? In exchange she would provide me with details of the smuggling group, and she, and she hoped her husband, would work for me as informers on Neo-Nazi organisations. (This sort of cool and selfish betrayal of fellow-nationals was astonishingly common among Austrians at that time, and we profited by it to no mean extent.)

Problem – what was this lovely creature up to? Factors –

(a) I knew, which she didn't, that the authorities were no longer interested in Karl.

(b) If I let her go she might well disappear with Karl. So what? We had no reason now to detain either.

(c) I would personally guarantee Karl's immunity from re-arrest, if he came back, use him and gain his confidence in return for his 'favour', and after some months inform him officially that he was no longer wanted.

(d) At the very least I would profit by getting details of the smuggling group, and I might also gain two informers.

Decision – let the lady go to Italy, after first checking the accuracy of her information about the smuggling group. It seemed, as it turned out, accurate. We knew the individuals and had suspected for some time that they were active.

Result – Karl and his wife returned. They did for a while provide a lot of information on what was never a very dangerous movement in Carintha, and are now well-known and well-respected citizens of a certain town in another part of Austria.

Love Conquers all

Jemima was petite, quite pretty, and spoke very good English with the faintest of accents. When taken off the train from Italy by my detachment at Arnoldstein, she was dressed in an ATS officer's uniform and had a large trunk full of luxury goods from Italy and an astonishing quantity of the more attractive items from RAOC Officers' shops. She had an automatic pistol, and an album with literally hundreds of names and addresses, and a number of photographs, of British serving officers. She had also – and this was the error which caused her arrest – otherwise excellent identity documents which the keen FS eye spotted as having been cunningly altered.

Jemima was undoubtedly a fake, but what was she and what was she up to? Interrogation eventually established that she was a Danish-German in origin, had been employed by one of the German welfare organisations in Italy, and after the surrender had decided, being an orphan, to live on her wits and her charms among the kindly and susceptible British. This, to judge by her address book, she had done with considerable success and was confirmed when she suggested to me with a shy smile that I would learn a lot more about her in bed, a suggestion which I, of course, had to refuse, albeit with regret. Why was she travelling to Austria? Well, British Forces in Italy were running down (true) and she had decided to graze new pastures. The pistol had been given to her by Captain so-and-so in Rome (established as true to that individual's acute embarrassment and subsequent courtmartial).

I decided in the end that Jemima was nothing more sinister than what she candidly confessed to be – an active and itinerate tart. She was prosecuted for illegal frontier-crossing and impersonating a British officer, and was given three months. She was then released as a Displaced Person and directed to one of the camps. Some months later I had a letter from the head of the American Counter Intelligence Corps in Salzburg saying that one of his men was about to marry a girl called Jemima who had quoted me as a referee and would I please comment! I gave CIC the full facts of her case and to my utter astonishment was visited a few weeks later by Jemima and her CIC husband on the way to Italy for their honeymoon. Jemima thanked me sweetly and demurely for all the help I had given her!

Man's Inhumanity

He was picked up walking along the railway line towards Italy, early one morning. He had no money, no documents, no food, absolutely nothing. He was in rags, had a serious speech defect, and whimpered with fear at the sight of a uniform.

He was the most pitiful object I had ever seen. His body was a mass of sores from – literally – the top of his dented skull to his toes. Part of his tongue was missing – hence his speech defect. Most of his teeth were broken. He was starving and was clearly ill. He was terrified of us.

Our obvious horror at his condition, and our sympathy, did eventually have an effect on his fear of authority, and we managed to give his story. A Jew, he had survived Auschwitz, only to be arrested by the Russians a few weeks after liberation on charges of 'anti-democratic propaganda'. He had apparently been foolish enough to criticise the new regime. After a year in detention 'a little better than Auschwitz', he had been released and was now making his way to

Palestine, he hoped. The Jewish underground travel organisation had got him as far as southern Germany, but they had panicked and scattered when approached by armed German police and he had carried on eastwards by himself. It was no good that German – and for that matter Austrian – police was now under Allied control. To him a police uniform spelt terror.

I had him sent to hospital, while I decided what to do with him. My duty was quite clear – he was an illegal frontier crosser and should therefore be subjected to the due processes of law, which would at best mean that he ended up in a German DP Camp, and at worst be deported to Germany. He was certainly no security risk – his appalling physical condition and the fact that he was clearly slightly unhinged mentally, bore testimony to the truth of his story.

What should I do? I had absolutely no hesitation in making my decision.

Trieste Security Office

In June 1950 I was posted to Trieste, where my first appointment was to run 'A' Liaison Section, which was simply a telephone interception service we had taken over from the Italians. This lasted about a year and was rather dull. I was then instructed to create a Screening & De-Briefing Centre, for the purpose of dealing with refugees from Yugoslavia or who came from other Eastern European countries through Yugoslavia. I had six British and American officers, who between them spoke all the Slav languages and Rumanian. Plus the usual secretaries/typists, mostly wives. Our job was to acquire military and industrial information,which we did. Masses of it.

After about a year of this, I was posted as 2 i/c to Trieste Security Office, which I fancy was one the best CI/Security organisations ever produced by the British Army made so by a succession of very good COs.

The strength averaged 5 officers and 50 men, mostly NCOs, and many successes during its years of existence. We recovered stolen top secret documents, ran agents into the Yugoslav Zone and occasionally into Yugoslavia proper, managed to keep tabs on many of the Yugoslav agents in Trieste and had a constant, guardedly friendly duel with Italian Military Intelligence. From time to time we did minor jobs for the MI6 people, affectionately known to us as the 'Chinese Laundry'.

I was eventually appointed CO.

When the great crisis arose over Trieste, with war looming between Italy and Yugoslavia, and both sides moving troops towards the territory, our families were hastily evacuated. I had NCOs travelling all over Northern Italy in civilian cars providing daily and detailed reports on ORBAT and movements. The Yugoslav Army was rather more difficult, but we didn't do too badly.

Somewhat later there was serious Italian Nationalist rioting in the city. The British-officered Venezia Giulia Police Force could not cope (troops had to be called out) and their information service broke down. I had NCOs, including National Servicemen, all over the city providing a steady stream of sitreps day and night, sometimes at considerable personal risk.

Then there was the time when a lieutenant-colonel of the Yugoslav UDBa (=Soviet KGB) defected and was smuggled into Trieste with his family. They were accommodated in our 'safe house' and debriefed over a longish period by various British and US agencies, including ourselves. For this purpose we would bring him to our office during the hours of darkness, in some disguise or other. On one occasion, for some reason which quite escapes me now, we had to dig him out of bed well after midnight, to question him on some matter which our masters considered urgent. He and his wife got it into their heads that this was to be his last journey, and I shall never forget that attractive lady, tears streaming down her face, clutching my hands and begging us to spare him. It took us some time to calm them down.

Eventually we were instructed to smuggle the family secretly aboard a ship in the port (if I remember rightly it was the 'Empire Windrush' which used to call occasionally at Trieste) and about one o'clock of a very dark morning we did just that. It was pure melodrama – the hushed exit from the safe house to a car, the journey through the silent streets, a whispered challenge at the dockside, and the final tearful farewells and gratitude from our charges when, in their comfortable cabin, they at last accepted the fact that they were safe.

When political settlement of the Trieste question was finally reached in late 1954, the city was to be handed over to Italy. As a preliminary, two Italian Military Intelligence Officers were sent in, in plain clothes, of course, to be attached to my unit in order to tie up security arrangements. After years of frequent virulently anti-British propaganda in Italy over the subject of Trieste, they were quite comically relieved to find that we were not ogres, and that we liked wine, good food and attractive women just as much as they did. They were excellent fellows, and we got on very well.

On the final days of evacuation and handover, there was to be a formal ceremony on the boulevard in front of the Excelsior Hotel, after which a destroyer tied up close alongside would take away the GOC and his aides, whle we few others would board an aircraft carrier, already fully loaded and waiting in another part of the port much further away. Italian Intelligence were worried by reports of a plot to assassinate General Winterton – never popular with the locals, during the ceremony, and I arranged that some of my best NCOs and another officer and myself would stay ashore, armed and in plain clothes, keeping in touch the while with our Italian opposite numbers and informants, and keeping a lookout out for various bad-hats. All this in and about the area of the hotel.

Overnight the weather became very bad indeed, and to my great relief the final handover ceremony was cancelled, whatever the real reason may have been. The aircraft carrier was not due to sail until midday, and we settled down to a celebratory glass. I had just downed a Cinzano when we received a message that because of the weather the carrier would weigh anchor early, and that if we were not aboard within thirty minutes we would be left behind. How I rounded up my chaps I do not know, but we squirmed our way through the still waiting and wind-battered crowds, found a reluctant taxi – the display of an automatic pistol and promise of a handsome financial reward made up the driver's mind for him and tore down in the heavily overloaded vehicle to the docks. More melodrama! The duty officer at the top of the gangway, seeing a disreputable taxi screech to a halt and disgorge five civilians, one of them tucking away a pistol, called out his marines (or was it ship's police?) and refused to let us aboard. Fortunately the Colonel (Int) appeared on deck at this point and we got on.

85 Port Security Section

Tony Thain

Revolvers on the Table

Three slightly similar scenes of Field Security life stick in my mind, each involving a revolver on the table in front of me. Two events occurred whilst I was a Sergeant in 85 Port Security Section, Algiers.

The first of these was around midnight on Christmas Eve 1942, when I was taken by a Gaullist contact to a flat in the City to meet one of the many plotters interested in the assassination, earlier that day, of Admiral Darlan. We sat for about half an hour on opposite sides of a table, his revolver on the table in front of him pointing at me, whilst I sought his explanation as to who might be responsible for the assassination carried out by the young Bonnier de la Chapelle, an explanation which I duly reported to my FSO later that night.

On the second occasion, a few days later, it was my own revolver on the table in front of me, sitting in the flat of Guy Menant, one of our regular Gaullist contacts, who, during the German occupation, had been responsible for the distribution of the underground newspaper 'Combat'. My task was to ensure that Guy Menant was not arrested that day or night by pro-Vichy elements of the Police who were using the pretext of Darlan's assassination as an excuse for rounding up their Gaullist opponents. Being young and unsophisticated in military affairs, I had not by then heard of a topic much spoken of in recent years and known as 'Rules of Engagement'. Fortunately for me, we were not visited, so no engagement materialised.

The third scene developed several years later when, having handed over the Indian Airborne FS Section in Karachi to an Indian officer, I was taking over command of 3 Field Security Section in Palestine in March 1947, a time when the Irgun Zweu Leumi and the Hagganah were devoting much of their energies to blowing up British military vehicles or buildings. The outgoing FSO assured me that he knew the hours between which the roads were dangerous and the hours during which the mines had been cleared in order to allow the Jewish farmers to tend their fields and orange groves. On this basis he took me in the FSO's Jeep from near Tel Aviv to a nightclub in Haifa, where we shared a table with his friend, a youngish woman who kept her handbag on the table in front of us with a small revolver protruding slightly from it. I learned then that she had been the mistress of the leading Arab drug smuggler in Palestine and that she kept the revolver handy, because she feared that the Arab was planning to kill her. Towards dawn the two FSOs returned to Tel Aviv encountering neither mines nor terrorists.

A strange assignment

Soon after my arrival on 8 November 1942 in Algiers with 85 Port Security Section, I was, as a Sergeant, given the strange assignment of trying to track down some Japanese, who, so it had been reported, were hiding in the Casbah. Dressed in civilian clothes which we had taken over from the Italian Armistice Commission whose villa became our residence, and accompanied by an Arab who had been recommended to us, I spent many hours in the Casbah in this search, but found no trace of anyone resembling Japanese or even Chinese. A difficult moment came when I was told that there might be such people in a Turkish Bath ("bain maure"), but I went in as a customer, leaving my clothes with an attendant. The awkward aspect of this was that in undressing and throughout the bathing I had to struggle to conceal my .38 revolver and my special FS Identity Card under the single large towel I was given. Nevertheless I did succeed and later emerged fully dressed, complete with revolver and Identity Card, but without any Japanese.

Norwegian Adventure – Memories of 89 FS Section

We flew along the coast, having seen it first at a point near Kristiansand, flew over Oslo fjord, over Oslo itself – which is small for a capital city, and looks small from the air – and landed at Gardemoen, about twenty five miles from Oslo.

We climbed out of the Stirling, and breathed air that that seemed clearer than English air. There were hills in the distance, slightly blue. Around the perimeter of the airfield were Heinkels and Messerschmidts, drawn up in rows, and some camouflaged in light colours for winter flying.

Germans were driving German transport on the runways. In one of these transports we travelled to the nearest railway station. There we were the subject of attention by the Norwegians. It was polite, aloof, Nordic attention, as opposed to Islamic attention or the attention of Latin peoples.

The language difficulty probably helped to keep us apart. Germans were much in evidence, but the white flags flown on their transport, and the predominent horse-drawn element of this latter, put them in a position inferior to us. There were no Germans travelling by train.

We were four FS, who de-trained at the Ostbanestasjon Oslo, that late afternoon the 12th of May, 1945. We did not know where the rest of our Section was: we had nineteen, all told, including our FSO, to account for. Two of us remained in charge of the baggage, two of us went off on a tour of Oslo, and a treasure hunt: the remainder of the troops marched through the streets of Oslo: It was a long hunt: the FS rendezvous which we had been given in Oslo was no longer available, it was barred, locked, 'stengt', as the Norwegians say, and not for British troops. Intelligence personnel met Intelligence personnel, and we contacted three members of 'Allied Forces in Norway' – Norwegian speakers – who had a lorry, which ran, just, on wood gas. With their aid, and the lorry's, we transferred our baggage to a large modern hotel. I learnt my first few words of Norwegian from the children: this was more 'it' and comparable with memories of the European countries during war time. Sweets and cigarettes came from our pockets.

A colleague and myself had a suite each that night. We were entertained until early in the morning by the staff and their friends – some of whom, I believe, were made spontaneously, as a result of their having heard English voices from within, as they passed outside in the street.

During the party – at which we had what had been once German food and drink – we observed a figure in a red silk dressing-gown moving a little somnambulently perhaps, between the tables. This turned out to be an important officer in British Army Intelligence in Oslo.

He told one of us that he had made contact with our FSO and Sergeant-major, and that no doubt we would be rejoining them in the morning. Having got in several hours sleep, after that little party, I was standing by my window, watching the Germans in a small depot of theirs across the street, when in walked my Sergeant-major. He had tracked us. And that was the end of the beginning.

There was lots to be done. We moved into an HQ that had been vacated by, amongst others, the SS. Our Oslo detachment was now seven strong. Eight others of our Section were known to be in Stavanger, three in Copenhagen, and one was still in England.

Oslo work was a succession of house searches, a period of Port Security, and a series of raids on German camps within a radius of a hundred miles or so, in a quest for arrestable personages, such as Gestapo members who' were in Wehrmacht or Luftwaffe umform and hiding in these 'lagers'. For most of these three types of work we co-operated with the Forces FS Sections. The camp raids were undertaken with our armed divisional troops: this, in fact was one of their chief tasks. They usually took place in the early morning, and it was a matter of leaving Oslo, in a small convoy of civilian cars and jeeps, containing Norwegians and British, the previous evening.

Rounding Up The Gestapo

In the early days shooting could be heard in Oslo every night. The Hjemme Styrke – Home Front – was dying to let off its guns, and any excuse was good enough. Our Sergeant-major was involved in a fine shooting affray, and we others were envious.

It concerned the editor of a German newspaper, who was reputed to have entered a flat. The attacking force shot away most of the flat's windows, one man advancing across the street behind a handcart: having stormed the stairs, the Norwegians and our Sergeant-major shot the door in.

There was no one in the flat. Fun had been had by all. Those men whom we did get – and the two Sections, co-operating, got quite a number of 'big-shots' – did not, to my knowledge, put up any resistance. Arrests included well-known Gestapo chiefs and members of SS Gericht.

89 FSS at Arnhem

Colin Ormiston
17 September 1944

89 Airborne Intelligence Section can retrace its origins back to the 28th June 1942, when it was officially formed. 89 Parachute Field Security Section (89 FSS), as it was, saw active service in Algeria, Tunisia, Sicily, Italy, Greece, France, Holland, Denmark and Norway. The actions of the Section on these operations have been inspirational and helped form a Section esprit de corps that is almost unique within the Intelligence Corps.

Four members of the section, under command of Captain Dunbar, were involved in the Bruneval raid to dismantle and return a Wurzberg precision radar dish. Captain Dunbar later died during the operation to seize Sicily. Members of 89 FSS were amongst the first ashore at Taranto, Italy, and almost definitely the first into the town. The Section's task was to search two enemy HQs, at the Miranda and Europa hotels, for documentation and booby-traps. The Section later commandeered a civilian car and drove through enemy territory to capture the German Consul at Bari. In 1944 many of the Section were transferred to the newly formed 317 Airborne FSS and saw action in Operation Overlord. But the best known of all actions fought by the Section was at Arnhem.

In 1944 the Section emplaned with a strength of 16; the Field

Security Officer (FSO), Captain (now Sir) John Killick, the Company Sergeant Major (CSM) (now Dr) Terence Armstrong and 14 NCOs. In the first wave the FSO and CSM landed with six NCOs. The FSO party the – FSO and four NCOs – moved towards the town with the lead Brigade HQ; the CSM and two NCOs remained with Divisional Tactical HQ to maintain contact between the FSO party, the other Brigade HQ, and the second wave. Communications with the FSO party were soon lost. The second wave arrived a day later with a further six NCOs bringing the total at Divisional HQ to nine.

On the 19th, Divisional HQ moved to Oosterbeek, a suburb of Arnhem, and the Section's security work began. The Section's initial tasks were to contact white list persons, search the homes of Dutch Nazis, compile lists of Dutch collaborators and interrogate Prisoners of War. The area was not clear of the enemy and sniping was continually in progress.

The Hartenstein Hotel, at Oosterbeek, became the Divisional HQ and was the next stop on the battlefield tour. The hotel now houses a museum dedicated to Operation Market Garden.

We sat and listened to a presentation which revealed that at this stage of the operation only the 2nd Battalion, with a Company from the 3rd Battalion. some engineer assets and some elements of 1st Parachute Brigade HQs had reached the vital Arnhem road bridge. All efforts by the rest of 1st Parachute Brigade and 1st Airlanding Brigade had failed to break through the strengthening German positions.

On the 20th the Divisional area contracted further; the security work ended. The Section became an infantry unit fighting in a defensive role. Gradually, the counter attacks of Panzer Grenadier/armoured battle groups of the 9th and 10th SS Panzer Divisions broke the survivors of the 1st Airborne Division into ever smaller groups.

Strenuous efforts were made by 4th Parachute Brigade to reach Colonel Frost's 2nd Battalion cut off in an ever-decreasing pocket around the northern end of the Arnhem road bridge.

With the 2nd Battalion were elements of 1st Parachute Brigade HQ which contained Captain John Killick and several members of 89 FSS. Sergeant Chambers, a man of whom it was said 'had 10,000 charmed lives', is reported to have stood, in the open, on the bridge, directing artillery fire at German targets, his reason being "that there was no one else to do it".

There were many other acts of heroism, but no one had the time to record them. On 21 September, after holding the northern end of the road bridge for more than four days with little food, water and ammunition, Colonel Frost's depleted battle group surrendered.

On the night of the 25/26th the Division was ordered to withdraw to the south bank of the Rhine. To carry out this withdrawal it was necessary to infiltrate through, or close to, enemy positions, and so the force was split up into parties of 10. The remaining members of the Section were all in the same party, but after walking into a German mortar position by mistake, found themselves split up. Only one member of the Section escaped from Arnhem unscathed; Corporal Zucker writes:

"Of the Section Philip Scarr and I were lucky. We were brought back from Arnhem and landed in Nijmegen . . . Philip and I went into the cellar of a Nunnery. It was chock-a-block with people. The nuns were praying. Philip Scarr went to sleep next to me. I found him dead the next morning. He was unmarked but his heart had given up from exhaustion. We were brought back to England and landed in Tilbury. There wasn't a dog to meet us. Only three of the Section returned. Our Field Security Section had virtually ceased to exist."

89 FSS at Arnhem 17 September 1944

Anon

The Section emplaned with a strength of 16 – the FSO the CSM and 14 NCOs. In the first wave (17 Sep) the FSO and 6 NCOs landed and moved off with a Bde HQ, 2 NCOs with another Bde HQ and the CSM with Tac Div HQ The job of the FSOs party was to go straight with its Bde into the town (the dropping and landing zones were about 8 miles outside it) and investigate the many targets there. The CSM was to remain at Div HQ and maintain contact between the FSOs party and the NCOs reporting in to Div from the other Bdes and the second wave.

By the following morning the CSM had been joined by the two NCOs from the other Bde. The FSOs party had presumably got into the town, but soon afterwards their Bde was cut off from the rest of the Div and no more was heard of the party – although it is known that the FSO at least had reached the main objective, the Rhine bridge. On the afternoon of the 18th, the second wave arrived, with them three NCOs with the third Bde HQ and three more with main Div HQ. These six found their way to Div HQ that evening or the following morning, after various individual adventures, and brought the total of FSS at Div HQ up to nine.

Div HQ had by now moved up into Oosterbeek a suburb of

Arnhem, and it was possible on the 19th to do some work. Houses of Dutch Nazis were searched and a quite comprehensive list of collaborators in the town was compiled from papers found. The area was still by no means clear of the enemy and sniping was continually in progress; a house-search party of two FS Cpls. was narrowly missed by a sniper, but one of them saw him and the other silenced him with a Sten burst. Owing to the small area held by the Div, work was scarce, so some FS kept themselves busy sorting the large number of documents found in houses that had been evacuated by the Germans. On the 20th the Div area was further contracted, and from that day on work was out of the question; Div HQ had to defend itself, and the whole area was continually mortared, shelled or sniped. Some FS were sent out on patrol during the days following.

Then on the night 25/26th the Div was ordered to withdraw to the South bank of the Rhine. To carry out this withdrawal from the Div HQ location it was necessary to infiltrate through or close to enemy positions, and so the force was split up into parties of 10 or so. The remaining FS were all in one party, but after walking into a German mortar position by mistake, found themselves split up. All reached the river ultimately, however, and all but two got across.

The casualties were severe. Of the nine at Div HQ, two were wounded rather severely and were evacuated to hospital on the spot, and two more reached the river bank but could not get across. Of the five that got across the river one died the following night and three had been wounded previously but were able to walk. So one man alone came out unscathed – and he was hardly surprised when he found that the seat of his pants was riddled with bullet holes.

89 FSS at Arnhem 17 September 1944

Anon

It was a fine, sunny morning in September 1944. We were in an old farmhouse in the Midlands, billet of Parachute Bde., Signal Section. After an early breakfast, three of us, the Bde. FS detachment for the op., got together and re-checked our inf. and briefing, examined our Stens, pistols, grenades, plastic, detonators, 24-hour ration packs, maps, and our other kit for the last time. 'Chutes had been fitted previously. At Bde. HQ we joined the trucks marked with our plane numbers (each FS NCO travelled in a different plane, in case of accidents). I joined the G3's stick, made up of clerks, batmen, 'I' staff, CMP, and other strays. After an hour's pause we moved off and were soon on the 'drome, joining our planes. A cup of tea and a bun, a snapshot taken by the American pilot, and then we got down to business. 'Chutes were fitted, men fitted with their various attachments, and we emplaned.

We took off in formation, and in a short time the sky seemed full of Dakotas and gliders as we flew southward over the English countryside. The sun was strong and our kit was bulky, the inside of the plane was like a hothouse. Most people were looking with interest out of the windows, one or two were reading 'Lilliput' or the morning paper. Our faces were red but cheerful and confident in the bright morning sun. As we left the English coastline, someone shouted "You've had it this time, chums", as if he had just realised this op. would not be cancelled like so many in the previous months.

We hooked up to our static lines, each man checking his neighbour. Fighter cover appeared all round us. Flak bursts appeared on the starboard side, our fighters dived. We saw no more flak till we reached the DZ (dropping zone). We had cleared the flooded part of Holland, and were trying to study the ground when the five minutes warning came up. It seemed no time before we heard 'action stations' and saw the red light. Captain T— was in the doorway studying the ground carefully, as usual. He was not excited, though he looked hot. He was too busy.

The green light popped on. We went out with a rush almost in a heap, a glance took in the sky full of flak, aircraft and parachutes. My chin took a crack from a badly-fitted Sten as my chute jerked open. I heard the rattle of weapons down below and the soft hum of lead in the air around me. But I was busy watching my landing. I landed gently, but had to wrestle to get out of my para harness. I shoved a magazine on my Sten. Then, being centre of the stick, waved the yellow signal to get the others together.

One of the battalions was hammering at Germans in a wood about sixty yards off, and plenty of stray slugs were floating around, so we kept well down and made our way to Bde HQ. We found this in a large hollow, getting organised. The two other FS men turned up and we set about interrogating prisoners, directing the searching, and sorting out captured documents. Most of them were ersatz SS clerks batmen, Dutch SS with a few real live SS men to stiffen them. We sorted out some of the units facing us, though they were mostly composite bodies, rushed together for the emergency. We moved half-a-mile on our way to a new position, as the battalions such as they were (very much below strength) were ready to advance, and our anti-tank guns had joined us.

We carried on with our duties till late afternoon, dealing with PoW and civilians. Following our instructions we left Bde. HQ at 17.30 for Div. HQ, accompanying and advising some RASC guarding a large batch of German PoW.

We moved slowly along battle-scarred roads and railway tracks. As darkness was coming on, I joined a passing jeep, reached Div. HQ half-an-hour later. I sought out the G.2 (I) and the Interrogation Sgt, and gave them details about the prisoners and information from Bde.

Then I joined the main body of our Section who were snatching a short two hours' sleep behind the 'I' tent.

When we rose, we immediately set about digging deep slit trenches for cover. In the morning the other two came in with bicycles and more material, having foraged around. They also had counter-intelligence information. The Sgt. Major allotted us jobs in pairs. Some went out to Bdes, some checked Div. HQ perimeter, as there were a lot of Dutch civilians floating around; I was given the job of sorting out all the documents, maps, etc., that came in from units and as a result of our own house searches. A German general had been shot up near his HQ, and we got a lot of useful documents. From house searches we had a pile of material covering collaboration movements throughout Holland with names and evidence.

During the first two days things seemed to go well. We had mortar barrages, airbursts from 88mm guns, and visits from German aircraft when our own aircraft were not supply dropping, but not on a large scale.

We heard nothing from the FSO and his party. We heard that the Bde and advanced elements they were with had gone straight into Arnhem and taken the vital bridge; we knew that they were still holding the bridge, but the Germans had brought in heavy forces and cut them off from us. People did get through to them a couple of times, but their news was not good. There was no news of the advanced corps of the Second Army, though the BBC news stated they had reached Nijmegen. Our own forces were pushing to get into Arnhem, but the Germans were superior in numbers, armament and supplies and looked like pushing us backward instead.

We carried on with our tasks; each day the going was rougher. Long mortar barrages came down on us, destroying most of our transport, blowing up ammunition and supplies. Movement was restricted.

Daily we had supply drops, daily the Germans increased their flak. Aircraft losses must have been heavy. We cheered the courage of the pilots who weaved their way through the wall of fire in their slow, heavy-laden aircraft; we were heartened by their bravery. Our perimeter was now very small and the Germans captured a great part of our supplies.

After four days we moved off about one hundred yards and dug in in defensive positions. We used corrugated metal sheets as cover from shrapnel. A German patrol broke through to within 150 yards of Div. HQ before being mopped up.

Snipers were never very far off. As soon as one was dealt with another would appear. We did infantry patrols, went out at night to help depleted anti-tank gun crews move their guns into different positions.

We stayed here until the end. The whole area around our slit trenches was flattened, houses became mere shells, trees scarred and leafless. The mortar barrages now carried on all day and all night long. German fighters buzzed around when our own planes were not there. Snipers picked us off as we moved around. Eight out of ten FS had wounds. Other units had so many casualties that the dead could not receive proper burial and were covered over in slit trenches.

We heard vaguely that a British Div. was fighting just south of the Lower Rhine (we could hear the roll of 25-pounder barrages) and we were always hopeful of seeing British tanks roll up to our relief. Our artillery started shelling the Germans from the other side of the river. They overlapped slightly when shelling the wood we were facing, and in ten minutes drove us out, more effectively than the Germans had done with their eight days of mortaring. We moved nearer to the house at Div. HQ, the basement of which was overcrowded with wounded.

At 1900 hours we were told that the Div. was about to evacuate. A British battalion was holding a strip of land south of the Lower Rhine, which we were to cross in boats. Fit men from each Div. party were to act as guides along the evacuation channel, a narrow corridor leading down to the river. I was the first guide on the way out of Div. HQ. The password was 'John Bull'.

I took up my position during a heavy barrage, feeling very much alone under a tree by a narrow lane, with the flashes and sounds of battle all around, and figures rushing about in the woods. I passed on all my parties to the next guide, also an FS man, on a crossroads covered by German MGs.

When the last party, the Div. Defence platoon, had moved on, I waited for several minutes, then, joined up with the next guide. We moved on, overtaking and passing several parties from other units, but saw no more of the HQ parties till we were near the river. The journey down was slow, it rained heavily, and we were drenched before we got very far.

Arcs of coloured tracer indicated the ferry point. Those who were seriously wounded on the road had little hope of attention and it was heart-rending to leave them, knowing that we could do nothing for them.

With three quarters of a mile to go we joined a huge, nightmarish sort of queue. The men huddled together for warmth and were silent. We moved along terribly slowly, but patiently, five deep. At times we lay down on the sodden ground. We were soaked through anyhow. With 400 yards to go to the river, we could hear the engines of the boats; but dawn was only about two hours off and there were hundreds behind us.

We got down near the beach, still moving along patiently. Something was going wrong with the boats. There were three only and they had been working all night long. One got shot up by German MG fire from both sides. After a time the two boats went off in the half-light and did not come back.

I thought of swimming across and went upstream with others so that I should not be carried too far down by the current. I was dog-tired and hungry, but took off some of my clothing: I watched others swimming. Daylight had come suddenly and the swimmers were being shot up in the water and on the far bank.

I saw one man go over the crest, but the others made excellent targets for German MG and rifle fire from both sides. The Germans were now less than 200 yards off to the east.

I dressed again and helped to build some sort of defensive position in our hollow with stones from the jetty. To leave the hollow we were in meant being shot up. We had one Bren there with one magazine. The Germans started mopping up from the east, popping mortar bombs into each dip in the ground. There were at least 400 men on the beach. Some had gone back towards the woods, but most were completely surrounded.

We were so closely packed that when those on the outskirts started to surrender, there was nothing for it but general surrender. I started to bury all my extra money, etc., and chucked my useless Sten into the river.

On surrendering, we were taken, about one hundred of us together, along the banks of the Rhine, still under fire from the west. Inside the nearest defensive positions we were asked to dump all remaining arms and ammunition.

We were marched off towards Arnhem through a badly hit residential area, strewn with unburied British and German dead, their faces blackened; past burnt-out Panzers and SP guns, broken down anti-tank guns.

German soldiers popped up out of the ruins of smart villas, threw us apples and Players taken from stray supply cannisters. As we neared Arnhem we passed many civilians with prams and carts loaded up with belongings. They carried white flags. A smartly-dressed German press photographer rolled up in a staffcar, his face wreathed in smiles, he started to take propaganda shots of us.

Until then we had been feeling depressed, dazed. We had most of us expected death, wounds, but never thought of capture; but at the sight of this Nazi we came out of our dejection with a bump. We shook him with V signs and grins, which made him dance and gestulate in a mad burst of temper. He contented himself with shots of the wounded.

We marched eastwards through Arnhem for $2^1/_2$ hours and were put into a garden near a Div. HQ. Fires were lit immediately with stray papers and maps that some people still had; last cups of tea were brewed. They took our names, rank and number, and we were searched, but not very thoroughly; they seemed keener to get field dressings and medical supplies than to get information. I retained my AB64 and someone else's fountain pen (not a good one – I swopped it soon after for two German cigarettes).

Our next stop was at Lutfen where we were confined to a large disused warehouse with a thin layer of dust on the floor. A German medical orderly came in, but gave up after about ten minutes, having no dressings left; most of the wounds were too serious for him to deal with. My friend the chief clerk, A/Q from Div. HQ, who had been shot through the nose, got a bit of sticking plaster; his wound had to heal up without treatment.

During our second day in captivity we had our first meal, some bread and sausage and some vile-tasting ersatz coffee. After another day we were given one third of a German loaf and a small piece of sausage – rations for three days. We were packed into box-cars. We had 57 in ours, too many for all of us to sit down at once; the doors were locked on us.

They were not unlocked for ten days, it took so long for us to reach Frankfurt-am-Main. Our water supply consisted of rain caught from the roof in steel helmets or water brought us by slave workers and passed through a narrow slot in the end of the box-car.

We stretched our cramped limbs at Frankfurt, received more ersatz coffee, bread and sausage. Our seven days of starvation had made several people very ill, but the only medical treatment they had was aspirin tablets.

The author spent the rest of the war in a PoW camp.

93 FSS Section in Luneburg

Norval Rodgers OBE, TD

On the 18th April 1945 a race took place between the Inns of Court Regiment and the 1st SAS Regiment as to which would enter Luneburg first. They came in from different directions and it is hard to name the winner. However, 93 FS Section deserved at least a place, as they arrived at the same time as the SAS, riding with them in the SAS jeeps by courtesy of that Regiment. The object was to enter the town in time to catch anyone on our Black List before he had time to run. Accordingly one of our first tasks was to restrict all movement by establishing check points on every road in and out.

On the day I have in mind, which was sometime later, four German Civil Police Officers apparently travelling from Hamburg were brought in to us from a check point because they were unable to produce the necessary travel permits. We had only 3 German speakers in the Section, all of whom were out on various assignments and so it fell to the unfortunate FSO, meaning me, to deal with these characters. They were all in proper German Civil Police uniform and at first sight looked as though they might be genuine cases of muddled papers. Two were middle aged and quite senior according to their badges of rank. The other two were younger and remained silent except when directly addressed.

After the usual preliminaries, I questioned the two senior officers as to who they were, where they were going and why. Also the reason why they had been unable to produce proper travel permits. In reply there followed a longwinded account of how there was a shortage of German Police in what was to become the American Zone and at the Americans request the Hamburg Police were posting these four south to act as reinforcements. This went on for quite a bit.

I began to address myself more and more to the one who seemed to be the most senior (he was certainly the fattest). Eventually the second officer (whom I will call the thin one) began to appear restive and he interrupted to ask why I did not address him, as he was the senior officer. The plump one did not dispute this and it came out that they were both of the same rank but one was of higher seniority than the other.

This was hardly surprising but there seemed to be a certain element of competition between the two, who began to argue between themselves. In the course of this I heard the word 'Standartenfuhrer', which was the SS rank equivalent to a full colonel in the Army. I must have blinked but hastened to follow up what appeared to be a verbal slip. In formal tones I addressed the thin one as "Herr Standartenfuhrer" and surprisingly he gave himself away by responding to it. This was an unexpected stroke of luck. The fat one now abandoned his police role and confessed to being a Standartenfuhrer likewise. The two juniors were mere Captains (like me) but they too turned out to be SS.

We now abandoned any pretence of requests from the Americans. These four men had shown themselves as Officers in the Allgemeine SS and further interrogation elicited that all of them had been members of Himmler's personal staff and travelled with him everywhere on the special train which he had made his mobile Headquarters.

Once the ice had been broken, information flowed more easily. Amongst other things I was surprised to learn that these SS Officers still had a strong sense of loyalty to Himmler. It seems he had taken a close interest in his personal staff and their families and an enduring bond had been forged. When I pointed out that Himmler had been responsible for inhuman cruelty to multitudes of innocent people, they replied simply that he took no pleasure in cruelty, he was merely impervious to it when he considered it necessary.

The real reason for the journey of these Germans now gradually came to light. They had hoped to make their way to Bavaria, where they thought there might still be a possibility of organising a resistance movement centred on Hitler's mountain resort at Berchtesgarten, known as the Eagle's Nest. They could not or would not accept that this plan had become unrealistic.

It did occur to me then that they might instead intend to join the Bundschuh (Bundschuh is the German word for a peasant's shoe or, in French, Sabot hence Sabotage). Bundschuh was intended to operate as a stay-behind underground organisation throughout most of Germany but it turned out to be a damp squib and, so far as I know, it never achieved anything of importance. If these four Germans had been pinning their hopes to it, they would have been in for a disappointment.

Finally, however, it was my duty to despatch under escort all four of these bogus policemen with my report on them to Camp 031, where there were the facilities, the personnel, the expertise and the time to carry out much more detailed and intensive interrogation than could be done at my level. And so we said good-bye to the four and never heard of them again.

93 FSS – Journey's End

An FSO's duties sometimes take unusual and unexpected forms

Norval Rodgers OBE, TD

World War II ended in Europe on VE Day 1945 but although the actual fighting stopped then, the surrender and movement of vast numbers of men, together with their weapons and transport had to be planned in detail but with little time in which to do it. That is where and when this story begins.

At the time 93 FS Section which I commanded was engaged in Luneburg interrogating suspects, mopping up undesirables, arresting those who were in what were known as automatic arrest categories, and carrying out security duties in general. Suddenly word came down from Headquarters 2nd Army that I was to report at once for special orders. Wondering what I had done wrong I duly reported and was told they were giving me an almost new Jeep and a first class driver. That was the good news. Then they gave me my instructions. That was the bad news.

Briefly, following on the German surrender to Montgomery, my task was to contact General Blumentritt's Army Group Headquarters at an agreed time and place and pass on to them the detailed plans that had been prepared setting out exactly how the surrender was to be carried out. On their part the Germans, who were understandably anxious to surrender to the British rather than to the oncoming Russians, were to provide information about their own forces and signify acceptance of the British plans. (They had little choice.) A great deal of detail was involved and everything had to be done in a great hurry.

I was to proceed at once to Hamburg and pick up some documents waiting for me there. Then, like a bat out of hell, on to a spot on the Hamburg–Kiel Road (map reference supplied) where some German Staff Officers would be waiting for me.

Accordingly I jumped into this nearly new Jeep, asked the driver his name and we were about to set forth, when we were suddenly recalled. They hadn't finished. Further documents were thrust on me and I was asked what I was waiting for, time was precious. Off once more and after calling in at the Hamburg HQ for the documents awaiting us there, we pointed the Jeep towards Kiel.

My driver was good, the Jeep was good and the road was straight and surprisingly well surfaced. Eventually, when we were beginning to suspect that the map reference we had been given might be wrong, we spotted three figures in the distance. They turned out to be the promised German Staff Officers, plus one very shabby, clapped out civilian car claiming to be a Mercedes-Benz. Everyone saluted everyone else most punctiliously, after a slight hesitation on both sides as to who should salute first. Then we examined each other's credentials and established that we all were who and what we ought to be. The Germans then asked politely whether they should lead the way and, bearing in mind that I did not know yet where we were going, I graciously agreed. Next, the Germans enquired what pace they should set and, after looking at their poor worn-out jalopy and then at our own Jeep, I said: "As fast as you like".

This was a mistake. That car had wings and although we just managed to keep it in sight on the main road, we had to drive flat out. When our guides turned left on to winding side roads, we lost them almost at once. They kindly waited for us several times with absolutely straight faces, for which they deserved full marks, but that seemed a very long ride.

At last, just as dark was falling, we reached a village which was evidently in use as a Military Headquarters. Here there was a wait for some telephoning to take place after which we were guided quite a distance to what even in the dark appeared a handsome and undamaged Landschloss or country house. This proved to be Journey's End. We were certainly in Schleswig-Holstein and just as certainly many, many miles from the nearest British soldier.

After my identity documents had once again been checked, my driver was taken away to thaw out – it was a chilly night – and I was led into the house and met by a senior Staff Officer, to whom I handed the documents I had brought with me. We discussed these briefly and he took them away for more detailed attention, after he had first warned me that the German papers he had been expecting from his own side were still incomplete and we should have to wait until all of them had arrived. He then left me in a large room which was obviously the library.

Time passed and a young officer who I later learned was ADC to General Blumentritt entered and invited me to join the Generals (there were three of them) for a glass of wine. I accepted gladly.

The ADC ushered me into a large drawing room where five or six people were sitting before a blazing log fire. I stood to attention and bowed in the German fashion to each one as I was introduced and I suddenly realised that my status had changed. I had become a guest, at any rate for the time being. A glass of really excellent wine was brought to me and there was a general air of relaxation. After all, hostilities had ceased.

The conversation, which had been interrupted when I entered, was now resumed. General Blumentritt, who was the Army Group Commander, his Chief of Staff and his General of Artillery were looking back over their campaign against the Russians and this made fascinating listening. After a while I was brought into the conversation to give a British view and in turn I was invited to ask questions. It was clear that the Germans still considered themselves as superior to the Russians in the field (which they almost certainly were) and I asked how they reconciled this with the Russians' success in driving the Germans back from the outskirts of Moscow all the way to Pomerania. General Blumentritt replied: "The answer is simply Mass". and he turned to his Chief of Staff and said "Give him the figures." The Chief of Staff knew these by heart and reeled off the number of divisions the Germans had facing the Russians – not only greatly outnumbered on paper but also each German division was so below strength that it had only the equivalent of a brigade to face a Russian division three times as strong. I had no answer to this.

General Blumentritt then put a number of questions to me about the British Army and (inter alia) enquired what the British Officers liked to drink. I told him that supplies were hard to get and we were forced to drink whatever we could lay our hands on. He twinkled at me and said "Well, at least you don't have to drink Französische Rotwein". This French red wine was notorious. It came from the worst dregs the French could find to supply the other ranks in the German Army. Although rather embarrassed I had to admit that we did in fact drink that if we could find nothing else. Blumentritt exploded: "What, that muck?" Oh, you poor fellows!".

The hours passed and at long last the papers we were all waiting for arrived and we were able to complete our business. After making my adieus and stowing away the documents the Germans had exchanged with me, I retrieved my driver, who had wisely filled up with German petrol and we found our old friends, the guides in their jalopy, ready to escort us once more. This time we proceeded at a much more sedate pace until we were back on the Hamburg–Kiel main road. Here we said farewell to our guides and this time we not only saluted each other, we shook hands.

There is little more to tell. After vainly trying to snatch a nap by the roadside – sitting huddled in an open Jeep proved too cold for that – we saw the dawn come up as we set off on our return journey. In due course we arrived back at HQ 2nd Army where I made my report and handed over my papers, which took some time. At length we were relesed to return to Luneburg. Neither my driver nor I had eaten anything for over 24 hours, nor had we drunk anything except (in my case) my glass of wine and we were happy to see the good lunch and drinks that 93 FS Section soon prepared for us. After this I said goodbye to my driver and we parted with regret. He was a good man and we would have been glad to keep him.

96 FSS
Activities of an FS Section on the Austro-Jugoslav Frontier

D.W. Sanders

The Eastern half of the Austro-Jugoslav frontier consists of densely wooded, almost mountainous country, most of which is inaccessible except on foot. Consequently, this is very favourable for illegal frontier activities such as smuggling, as well as being the haunt of political couriers and armed bands. These bands consist of persons exiled from Jugoslavia, who periodically cross the frontier, to and fro, with the object of distributing anti-Tito propaganda, killing Jugoslav soldiers and looting and disturbing civilians on both sides of the frontier.

Since 1946 this area has harboured a famous gang of of almost 25 desperate armed ruffians, claiming to be opponents of the present regime in Jugoslavia, who lived by looting and robbing on both sides of the frontier. This gang is also known to have been, either directly or indirectly, responsible for the death of at least ten persons in Austria and Jugoslavia.

The political implications of the presence of such a band in Austria were very serious for the British Authorities, as the Jugoslavs, seeing that they used the area as a base for their raids into Jugoslavia, alleged that the British were not only tolerating their presence, but even assisting and arming them. This was, of course, completely untrue.

Since they arrived in the area, strong patrols of British Infantry, accompanied by Austrian police, were constantly searching for them. The difficulties encountered however were very great, as apart from the close nature of the country which afforded excellent cover, it was difficult to obtain information of the gang's whereabouts in sufficient time to effect their capture.

Austrian farmers, who were forced at the point of a machine-pistol to give shelter to the band, could not in any case inform the authorities until the band had left a particular farm, owing to the poor communications and widely scattered police stations in this area. It was therefore often a matter of several hours before the authorities received information as to where the gang had been in hiding. It need hardly to be said that most of the farmers were terrified of giving information against the gang, who had threatened them with sudden death if they did so. It will be seen that the capture of such a gang of desperados deepened for the most part on luck and that any detailed plan of capture was quite useless.

Tension in the frontier area increased owing to the continued presence of the gang and their extensive raids into Jugoslav territory. Communist propaganda fanned the flames. The Jugoslav frontier guards became increasingly suspicious of any person they saw in the immediate vicinity of the frontier and, as a result, a number of innocent persons lost their lives.

As has been seen, only the merest chance would bring a British patrol into contact with the gang. In July of 1947 a brush between the gang and a patrol of Field Security Personnel and Austrian police took place in which shots were exchanged and two of the gang were wounded. However, they managed to escape across the frontier into Jugoslavia.

It was decided that the time had come for subtle methods to be adopted. An agent of the section became a member of the gang (they were not particular about who joined them) and was able to give valuable information about personalities, weapons and activities. On numerous occasions the agent informed us where the gang would be at a certain time, but it was not until late August that our efforts at capturing them met with success.

In the early hours of the morning our agent, obviously having sampled far too much of the local 'schnaps', informed us that the gang were indulging in an orgy in a nearby hut. Appetites whetted by this information, the whole section, armed with Sten guns, set off. The meeting place was surrounded and on entering we were met with the sight of some ten persons, both male and female, some in rather compromising positions. The male members of the party were armed with various weapons, but owing to their inebriated condition they were unable to use them. All our captives were transported to the local jail, where they were given ample time to recover from their carousals.

The result of this capture helped greatly in improving the relationship between British and Jugoslavs in this area of the frontier. It appeared that the Jugoslavs had themselves captured most of the remainder of the gang during one of their raids in June. The OC and two of the Section NCOs who later made an unpremeditated visit to Marburg (Mariborg, Jugoslavia), were especially well treated and congratulated by the Jugoslav Authorities when it was learned that they had personally been responsible for the apprehension of the bandits.

As a direct result of the eclipse of this gang, the frontier area immediately became quiet, the civil population was allowed to carry on with its lawful activities in peace again and relations between troops of both nations on the frontier improved, despite the continued propaganda which was installed into the Jugoslav soldiers concerning the intentions of the British. The charges which the Jugoslavs had in the past levelled against the British for assisting, arming and using this gang for its own ends had been finally answered.

258 FSS

Maurice Davie

On the 21st November 1942, Captain Scott, myself and a driver set out on a Security Mission into the interior of the Egyptian desert, our destination was Siwa Oasis, far from any inhabited locality and about 200 miles south-west of the port of Mersa Matruh. From the time I had landed in Egypt I had a strong desire to visit this ancient landmark, and now the opportunity presented itself.

Prior to this forthcoming adventure I had seen only the coastal strip of desert which at certain times of the year was aglow with the colours of short-lived flowers, and the 'camel thorn' species of heather. Scattered along this coastal strip were Bedouin tribes and encampments, adding at least, some life to the only desert I know. However, I was soon to become aware that my impressions were misleading and that a desert could become a tough proposition even in the days of modern transportation.

We left our base at Alexandria and headed due west along the coast road. It was our intention to reach Mersa Matruh by evening and stay there that night. The first stage of our journey was accomplished without incident. During the journey we had a magnificent view of the El Alamein battlefield which was by this time deserted. The devastation rendered on the enemy was colossal. Every few yards there was a burnt out tank, a truck or an overturned aircraft and piles of empty shell cases and unused ammunition. This was a sight I should never forget.

Travelling in a staff car can become a tiresome business after a whole day over a rough shell-torn road and it was therefore not surprising that we were tired and hungry. A meal and a hot drink was produced and we finally bedded down in an old derelict building. A good sleep was denied to us due to the howling wind outside and the masses of fleas and bugs inside the building.

Early the following day we started out and we hoped to reach Siwa by nightfall. Before us lay over 200 miles of rough, rocky desert, devoid I was told, of any form of communication with the outside world. The thoughts of this latter prospect held no immediate horrors for us, on the contrary, we were looking forward to the trip. To my surprise the going was fairly good, and for the first 20 miles we were travelling on a good road. The scenery at this stage could not be compared with the Lake District, the wooded hills of Surrey or the lovely green meadows of Leicestershire, but wild as it was, this landscape had its attractions.

After a time the road gave place to a rough track which varied little from the surrounding countryside, and it was then that I began to realise the real meaning of the word 'desert'. There was nothing but rough stones and building sand as far as the eye could see, yet here we were endeavouring to conquer the elements on this crazy but essential errand.

We carried on with this monotonous journey until sundown when a strange sight caught our gaze. Two figures were stalking towards us. They could not possibly be Germans or Italians we argued, for they had been driven out of the area over three weeks previously. The two men staggered to the car, and what a weird spectacle they presented. Both were bearded and exhausted. Slowly and between signs of utmost relief they unfolded their story. They were English all right. One was a native of Liverpool and the other originated from Birmingham.

Their story was one of horror, more akin to fiction than stark reality. Ten days earlier they had become detached from their unit when their truck had developed trouble, and they had wandered aimlessly through the wilderness for over 70 miles. There only food was a tin of marmalade found in an old dump, and for drinking water they had drained radiators of smashed vehicles. Little wonder at their condition, poor devils. They related how they had argued about which direction to follow and had almost come to blows, and at times they had prayed in their last struggle to control their senses.

They were made comfortable in the back of the car and eventually we took them along with us, but we made a terrible mistake. The wisest course would have been to return them to Mersa Matruh and lodge them with one of the small detachments there, but as our journey was one of emergency, we had no alternative but to go forward. In any case we were almost certain of getting extra supplies at Siwa.

The sun began to set and it quickly became dark. The car was stopped and we made a fire, brewed tea and cooked some tinned food. Never before had I seen anyone devour and enjoy a meal as these two did. Relief was plainly visible on their weather-beaten, sun-tanned faces. Chattering over the last mug of tea, the last remaining gruesome details were told and they thanked us over and over again for such a timely rescue.

We continued our journey, the moon had fortunately risen by this time and it helped considerably. I soon began to take a dislike to travelling under these conditions and I think the others shared my views. We were completely isolated except for the guidance of the stars and our compass and of course we trusted the Captain's intimate knowledge of this route to get us to our destination.

Long, weird black shadows emanating from rocky hillocks and sand dunes gave the appearance of monsters from a strange world awaiting their chance to close in. This illusion was uncanny, but in reality all around us was a vast expanse of barren earth as dead and as cold as the moon itself.

In whiling away the time I continually and mentally recorded the compass bearing. A general south-westerly direction that varied slightly. At this stage the Captain informed us that we had covered about 120 miles since leaving Mersa Metruh and that we had a further 80 miles to cover. The unexpected then happened, the gears of the car would not engage and we came to a halt. A quick inspection confirmed our fears. Serious trouble had developed in the transmission or gear box, and it appeared to us that nothing could be done until morning when daylight would be an asset.

It was already becoming quite cold, but this was very characteristic of the desert at night, and as a form of protection we hung our groundsheets around the outside of the car. The five of us then crept inside and settled down as best we could. During the next four hours the temperature dropped about 50°. The change was so sudden that it was almost terrifying.

Dawn came after what seemed an eternity, and we then realised what a desparate situation confronted us. The horizon appeared to be so distant that it seemed to become part of the pale blue sky. I made a survey with the binoculars, and in every direction there was nothing but a sea of rock and sand. Not a living thing could be seen, not even a fly.

After a quick breakfast of bully beef and biscuits we carried out a further inspection of the car and made several adjustments without success. We were stranded, and all about us was a deep silence such as I had never experienced before. To make matters worse we had strayed from the regular track, by how far we did not know.

We discussed the grim situation and it was decided that the Captain and I who were the fittest of the party should walk back to Mersa Matruh and gain assistance, leaving the driver and the two men with the car and the bulk of the food and water. It was a colossal risk but we had no choice.

Carefully we made preparations packing only the bare essentials with a minimum amount of rations. We also carried pistols and two pockets full of ammunition and a map given to us by one of the two men. This map, according to our calculations, fell short of this area by about 40 miles, which meant having to walk blindly for this distance.

The big moment came and we said "Goodbye".

Neither of us dared to entertain the idea of failure. I set the compass and we started our long walk. The sun rose higher and it became unbearably hot. After an hour a track came into view and then like a mirage there appeared an old signpost pointing in a northerly direction with the single word 'Mersa' painted thereon. I had been deceived, for here in the middle of nowhere was a signpost. With renewed confidence we strode forward.

After a time I saw two unusual mounds of earth or concrete in the distance. I was mystified, but the Captain obviously knew what they meant. Time did not allow for sightseeing so we carried on in silence. Later I realised that this was Bir El Nubs, translated as 'Half way well' or "Half well'. The well was apparently situated halfway between Mersa and Siwa. In the past it must have been a very important watering place for the Bedouins. The track became very rough at times with an over abundance of loose stones, and naturally this was hindering our progress and playing havoc with our feet.

Later in the day the track led off in a north-westerly direction. This created confusion. There appeared to be no reason for it and we argued, but eventually we decided to follow the track. It continued in the same direction for about a mile, and then, convinced we had made a mistake, altered our course to north-east again. The nature of the ground became peculiar. With each step we took, we sank about two inches, when our feet penetrated a thin crust similar to the ice on a frozen meadow. There were no other footmarks or wheel tracks or any signs of any living thing having ever been in that locality before. Could this be real or were we indeed in a strange world. We often came across this type of terrain in these wadis (valleys). They may have been dried up salt lakes.

Our clumsy packs began to weigh heavy and our feet and legs felt like lead. We halted. Foolishly we drank from our small ration of water. However, it revived us and our aches and pains temporarily vanished. Time was still precious and we resumed our trek. On across this rocky plain we struggled until even the sound of our footsteps became unbearable.

Despite the lack of scenery, the setting sun with its rich shades of red, orange, blue, purple and green did its best to break the monotony. In a few minutes it became dark and we settled down in a tiny wadi for the night. It was not quite as cold as the previous night and we soon made ourselves comfortable.

The night passed without incident. I was fully awake when the first streaks of dawn appeared and I promptly aroused the Captain. A quick meal and we were off again. The coolness of the early morning air gave us additional vigour and more valuable miles were convered.

A rough check of our position indicated that within the next few hours the southern limits of the map would come into use. Only then we would be certain where we were.

In the distance hills and valleys came into view. This was a welcome change from the first monotonous plain we had travelled. By about 8am the dazzling sun like a fiendish monster, began to extend its searching rays earthwards and to bleach still further, the whitened earth. It also succeeded in tormenting and hindering us in our errand of mercy. Amongst the rock and sand we found some disused infantry positions with a maze of shallow trenches but there were no signs of recent occupation. However it was a remote sign of civilisation, and any sign of that was welcome. We searched the area for water but in vain, instead we located a huge dump of abandoned drums of oil. In desparation we set fire to this oil hoping to create a smoke signal big enough to be seen perhaps by a patrolling aircraft. The fire gained in intensity and there were loud reports as some of the drums exploded, but as we walked away we felt that the fire served no useful purpose.

The Captain drew my attention to a track which suddenly appeared to our left. This was obviously the track to Mersa and we must have walked parallel to it for miles. On the skyline was yet another welcome sight. There were the familiar mounds of earth which denoted a well, and in our eagerness we bounded forward. Our water was getting low. On reaching the spot I saw a palm tree about four feet high and as green and healthy as a patch of English grass, standing alone in this vast desert. It was an amazing sight, yet alongside was another spectacle. In the well were thousands of gallons of cool, still water. On closer inspection it had the same colour as milky cocoa, but it tasted quite alright. We boiled some of the water over a fire made of brushwood and filtered it through a handkerchief, it was quite palatable.

A further check with compass and map and this revealed our true position as Bir El Fouad, the well, named after ex-King Fouad of Egypt. Water bottles were filled, we washed, and were completely refreshed, this was a wondrful feeling.

The Captain suggested trying to reach another landmark by nightfall, a spot marked on the map as the 'Government Wells'. We struggled on between short halts for a further six hours until our movements could hardly be described as walking. The daylight faded on yet another day. A few miles ahead of us on the skyline was the ever welcome sight of two mounds of earth. Our aches and pains miraculously disappeared as we hurried forward and confirmed the presence of water. Another meagre ration of cheese and biscuits followed by more brackish water sufficed our evening meal after which we snatched a few hours rest.

Dawn broke and preparations were made for the third day of this exhausting struggle against time. My thoughts wandered to the three men we had left behind, how were they facing this ordeal? Had they in their desperation wandered away. The night's rest had little effect, our feet were becoming badly blistered due to the rough terrain and Army boots, and our throats were developing a dry soreness obviously caused by drinking well-water.

The next objective was Bir Istabel, ten miles further on, it took four hours. This well was dry. Naturally it was a disappointment but this was to be expected. The wells would either be dry, filled in, or poisoned as we approached the battlefield.

Bir El Hilw was next on our list, translated, it was described as the 'Sweet Well' and this was a further ten miles away. In some of the small wadis and salt flats were deep ruts caused at one time by the wheels of heavy vehicles and to my astonishment they contained just a small amount of water. Perhaps it was dew or the remains of a slight rain shower which was a rarity in those parts. We had to lap this water like dogs but at least it was drinkable.

The noise of an aircraft passing overhead attracted our attention and our first thought was that our earlier smoke signals had been seen. It was amusing the way we waved towels and flashed an old piece of broken mirror trying to send an SOS, but the aircraft flew on.

Bir El Hilw was reached about 5pm and this well was also dry. This was a serious situation. The soreness in my throat in the meantime had grown worse, so had Captain Scott's, and we found it very difficult to eat dry food. We tried frying some cheese in an old tomato tin and we managed to consume a little food that way. After walking for a few hours we stopped and rested. The Captain suggested we should start very early the next day and to continue to walk as far as it was humanly possible.

At 2.30am we were on our way. The air was delightfully cool and fresh and to our surprise we made good progress. Eight miles were covered in three hours. Suddenly our thoughs were shattered by the barking of a dog; we listened. Again the barking reached us clear in the distance. We peered into the darkness, and there silhouetted against the sky were the figures of camels, goats and donkeys, and alongside a tribe of Bedouins. They were stirred to action by our presence, and came towards us asking "Mein Inter"? (Who are you?), Captain Scott replied, "English".

We were conducted to their camp and majestically they laid out a multi-coloured carpet for us to sit on. In fluent Arabic the Captain disclosed our plight, and we were presented with an assortment of foods, dried grass, dates and beautiful, clear water. With the advancing dawn I could distinguish their dark, clear-cut features, pointed chins, short beards and jet black, deep-set eyes. Truly 'Masters of the desert'. They needed neither map, compass, sign post nor motor vehicle.

Water bottles were replenished and with a good supply of dried dates in our packs we thanked them and bade farewell. Daylight came and by this time we had reached Bir El Gellaz. This was an oasis comprising about six acres of semi-fertile land where birds and insects abounded. The water too, in my opinion was equal to that of the finest French champagne. Perhaps the word 'Gellaz' meant clear, I did not know.

All through that day we stumbled on. My feet were in a terrible state. All my toe nails had blackened and some had fallen out, and when darkness finally came we both fell to the ground exhausted beyond all imagination. We slept soundly.

The next morning, after about three hours walking, Captain Scott drew my attention to what appeared to be an Army truck in the distance. I used the binoculars and saw several figures moving nearby. With a supreme effort we pushed forward and on reaching them we were greeted with startled looks by an officer and eight other ranks. They boarded the truck and hurriedly left, but they promised assistance. We waited.

Two hours later a military ambulance arrived and two of us were conveyed at break-neck speed to Mersa Matruh, and there we related our story to a Colonel of the Unit. The camp soon hummed with activity. Medical attention and food was given to us and within the space of an hour the Captain and I were in a well organised convoy of ambulances, wireless trucks, and a supply truck, wending our way back to the stranded car which we reached by daybreak the following day.

There in the centre of that cruel wilderness we found our weary companions. Words could hardly describe their emotions. They shed tears of joy and were overcome with relief.

Little more remains to be told of this story, Another long, tiring, rough ride back to the coast and civilisation, and it was all over. The two men were placed in a transit camp and the car was towed back to Alexandria. We had however, exchanged names and adresses and had said 'Goodbye', hoping that one day we all might meet again.

I wonder!

The World War II Years of Defence Security Office, Alexandria

Maurice Davie

In order to appreciate the functions of this office, which have naturally varied in scope and direction from the outbreak of war till the present time, it is necessary to survey very broadly the different security problems with ADSO Alex has had to face during the past $5^1/_2$ years.

In the early days, 1939/40, the activities of this office were practically those of taking measures to prevent outbreaks of sabotage in the docks, petrol areas, oil installations, water and electrical, etc. Early in 1940, before the entry of Italy into the war, an internment programme was embarked upon and various controls, travel access to docks, etc, were gradually but effectually tightened up. The first cases of enemy sabotage occured in April, but these were fortunately discovered in time to stop any serious consequences.

As the year passed, signs of enemy espionage activity were noted, the leakage of information being traced to a group of Italian women who were in contact with the Jap Consulate, when Italy joined forces with Germany. A large number of Italians were interned and this programme was continued during the next years.

Early in 1941, efforts were made to penetrate the Jap Consulate, which was considered to be the principal source of Axis information. It is worth noting that, whereas previously investigations of this nature were carried out largely through the Alexandria City Police, an independent organisation had by this time been set up for the purpose by ADSO Alex, later in the year with the removal of the Japanese Consulate from Egypt, there was a noteworthy falling off in enemy espionage activities although these were taken over by non-interned Italians and also Copts., and propaganda remained a serious threat. Sabotage was surprisingly non-existent, this was surprising in view of the increased difficulties entailed in maintaining a high standard of security in the extensive dock area, an obvious field for enemy agents.

Meanwile reports from the usual, normal routine investigation and control work of this office, a number of other varied problems had to be faced. The white Russian community constituted a threat to security during the period of the war. Local Egyptian opinion both in the Army and amongst civilians, swung this way and that with the tide of the desert battles, and needed constant watching. Greek, Czech, Polish, Free French and later other Allied forces, were stationed in Alex area and although they did not at the time cause any actual trouble, there were reactions to events taking place in their respective countries which had to be carefully gauged.

In 1942 it became evident that the Axis were employing more and more local subjects, Egyptians, and other residents of Alex, as espionage agents and certain isolated cases of espionage did in fact take place. In the summer of 1942, two Italian sailors were landed near the harbour by two-men submarine for sabotage purposes, one man was captured shortly after the autumn, but not before damage had been done. It was noted that Bedouin agents were being employed to travel back and forth behind the Alamein line, which was so close to Alex. An effective security measure was taken when a Pass and Permit was established in this office, and at the end of the year, after our success at at Alamein, security was considered to be very satisfactory.

In 1943, as the battle zone drew further and further away from Alex, the danger of enemy espionage and sabotage was lessened, although a serious case of the latter, the destruction of the 'SS Clan Macindoe' occured in mid-summer. generally speaking however the functions of this office were gradually transformed as time went on, local politcs and political activities amongst Allied forces stationed here, Alex came to the fore. For a year past, evidence had slowly accumulated on an illegal, Greek, Leftist organisation which was working to overthrow the Greek government, and then long expected troubles came to a head when Anarchist elements on board HMS Hierax mutinied in Alex harbour on the 16th July. Repercussions of the event continued to be felt for several months afterwards, and the security problem was no mean one.

In 1944 (April) this event was followed by a mutiny in the 1st Greek Brigade which again entailed a large amount of work for this office, in the way of arrests, interrogations, etc., later in the year, the Palestine problem began to make itself felt in this area, and investigations were accordingly made into Zionist and more extreme activities. In September of this year a report to this effect that certain Jewish extremist members of the Jewish forces were planning to sabotage the Pan Arab conference was investigated.

In Cairo with 259 Section

John Braun OBE

After six months in the desert I was sent on a Field Security course in Cairo in February 1941, then posted to 259 Section, operating from the Bishop's House, Garden City, which also served as the billet for the section and HQ Section.

The run of the mill security work in 1941 was the same as that in any city, except that Cairo was larger, noisier, more densely populated and more cosmopolitan than most. Flashing lights abounded; the most blatant failure to observe the blackout occurred in the rooftop apartment of the Head of the Cairo City Police; the search for escaped Italian PoWs was a regular chore; and the campaign against careless talk unremitting. For the last of these tasks, one of our ploys was to drop visiting cards in bars, the many open-air cinemas, cabarets and other places of entertainment frequented by the troops. We had three different cards, with the following messages printed in capitals. REMEMBER WHAT JONAH TOLD THE WHALE,"IF YOU HAD KEPT YOUR DAMN MOUTH SHUT, ALL THIS WOULD NEVER HAVE HAPPENED"; "PLEASE DO NOT TALK SHOP: ONLY ENEMY AGENTS WANT TO HEAR IT"; and "KATHERINE PARR NEVER GOSSIPED: SHE KEPT HER HEAD". My favourite was the visiting card inscribed: "U.R. Requested" with "2 Refrain from Careless Talk" in one corner and "Tel. No 1" in the other.

The intermittent searches for illicit W/T sets were enhanced by close cooperation with the Egyptian Police, one of whose officers, Yehia Mahmoud, was a good companion and competent operator who would invite members of the Section to join him in his normal police work. I once accompanied him on a midnight ambush of a cat burglar, who dropped from the low roof of a building into the waiting arms of four policemen and then howled all the way to the Muski Caracol, where he was charged. Another extra-curricular jaunt was a raid with one of Russell Pasha's squads on an opium den in the heart of the city.

The loss of Army Paybooks (AB64s) was giving rise to concern during this 1941 period. It was thought that enemy agents might be making use of them and infiltrating in spurious uniforms. Somebody had the bright idea of a raid on the Berka, the brothel area behind Ezbekieh Gardens. It was not a pleasant exercise, trailing in and out of dozens of squalid rooms but 259 Section came away with a haul of over 100 books which had been left with the trusting 'girls' as security for future payment.

Another type of raid was into the homes of 'escaped' internees. There was a large Italian, and therefore enemy, population in Cairo and the males of military age had been rounded up and interned by the Egyptian government. It was common knowledge that the Egyptian guards were regularly bribed to let out their charges to spend the weekends with their wives and sweethearts. The OC received orders to round up and return to prison any such escapees. So, one weekend the whole section, together with a squad of Egyptian police, descended on Shoubra for a dawn raid. We turfed out any number of internees on unofficial 'leave', hauling them out of their shared beds to the outraged indignation and venom of the womenfolk, who screamed abuse at us and clawed at us with their sharp finger nails.

Two episodes linger in the memory for their Keystone Kops elements. Shortly after I had joined the section, an Embassy tip-off came in of the alleged theft of some secret aircraft plans, which were about to be taken to an address in Alexandria. We alerted the section there and having missed the suspect, who was travelling by train, at our end, left it to them to cover the Alex station. Artie Shaw drove four of us at breakneck speed along the desert road but we were too late. The other section had apprehended the suspect and were interrogating him at their headquarters. When we arrived there, the interrogation had been completed and the suspect was about to be released. Our quarry had been a very frightened 15-year old lad who was clutching the evidence, a set of plans for making a wooden, elastic-driven model areoplane.

One of our star turns was Cyril Benjamin, who spent nearly all his time in a civilian suit and tarboosh, although he could not speak a word of Arabic. He was an expert observer and was able to turn his observations to good account because he was also an artist and could quickly sketch suspects. He had struck up an acquaintanceship with an RASC corporal draughtsman who worked in one of the map depots whom he had persuaded that money could be made from a surreptitious deal in military maps. Matters moved forward and we were all set to bring off a great coup by arresting the errant corporal as he handed over a batch of maps for a fistful of piastres. I was then sergeant in charge of the office, compiling the weekly reports and liaising with the CMP and Defence Security Office (Egypt). On the very eve of our triumph-to-be I received a call from my opposite number in SIB Bab-el-Hadid to the effect that one of their stooges had been posing as a mapper and was about to entrap a grubby little man who had been negotiating to buy some maps from him. As he was a civilian, this was an FS matter, so would 259 Section cooperate in bringing the suspect to justice? "Was this fellow an artist?" I asked. "Apparently". "Did he usually wear a blue suit and a tarboosh?" "Yes". "Oh" said I, "I think the wires have become crossed". And that was the end of that.

A more positive episode was action following reports of the theft of arms and ammunition from the Tura caves ordnance depot. Apart from the reports which filtered in and had to be strung together on my cumbersome Oliver, I had not been directly involved. The day came for the 'sting', when I was called in to augment the team, which included SIB and Egyptian police, all under the command of the FS Commandant, Major Robin Wordsworth. The ploy was to plant bait, in the form of decoy arms and ammunition in boxes, in a disused garage in the Bulaq district. Contact was to be made with the traffickers, through some shady intermediary, and the principal lured to the garage for an inspection of the haul. At that point they would be arrested. I donned my seedy grey civilian suit, thrusting my Smith & Wesson 38 into my trousers pocket. I was given my post at the end of the street in which the garage was located, with instructions to stop any speeding car or running miscreants who may have escaped the close net of watchers. A colleague was posted at the other end of the street in a similar longstop position. I was half-concealed in a doorway, sometimes lounging, sometimes resting on a brick cornerpost, becoming more and more conscious of the bruise being made on my hip by the wretched pistol. The sting had started in early evening and the brief dusk had become darkness when nearly three hours after taking post, I heard Wordsworth shouting. I leapt from my cover and went up to him. He asked me what on earth I was doing there and why I had not reported in with the rest of the section. I replied that I had heard nothing and seen nothing but was carrying out the last order which was to be on the alert. I asked when the business was to finish and was told that it was all over half an hour ago. "Had it gone well" I asked. "Well enough," he replied, "but we didn't catch the man we were after because he didn't turn up". I learned later that one civilian Army lance corporal had been arrested.

Some of the repercussions did cause a measure of excitement. As part of the follow-up, a number of raids on cabarets were organised, soldiers and civilians being involved in the arms racket. These were carried out by CMP and FS and I took part in a raid on the Sweet Melody Cabaret in Sharia Emad el Din. This was a very seedy joint, raucous, smoke-filled, with mixed troops, largely Australians, who were more drunk than the New Zealanders, who were making more noise. The main suspect was the drummer in the band. Following the advice of an old hand, I planted myself near an exit, with my back to the wall, as a couple of burly MPs marched straight up to the band and seized the drummer, carting him off the platform. The band stopped playing and they all started screaming and waving their instruments at the intruders. The military patrons, denied of their music, created hell. Bottles began to fly, one crashing just above my head, covering me with splinters of glass. As the MPs, their quarry secured, took him off to their truck, the FS support dived smartly out through another exit.

On Pearl Harbour Day, 7th December 1941, I left 259 Section to be commissioned as an IO in DSO (Egypt), part of Security Intelligence Middle East (SIME) and so was not with the section when it had its greatest and well publicized success in catching the German spy, Johannes Eppler who, with his W/T operator, Sandy, was trailed from their arrival in Upper Egypt by Capt Sansom, the OC. He and the Section kept them under observation while they operated in Cairo from the dahabieh of Hikmet Fahmy, the belly dancer at the KitKat Cabaret. From time to time I would join in a stake-out, dressed as ever in the same old grey suit. The KitKat was under constant surveillance, which was a great attraction for 259 in the warm summer evenings of 1942 while the Alamein build-up was proceeding. The link between the two events was the discovery in the desert, during a July raid, of the code based on Daphne Du Maurier's novel 'Rebecca'. One evening, four of us were sitting in the garden of the KitKat, waiting for the open air cabaret to begin, when four huge Egyptians walked in, making for a nearby table. Smurthwaite, who was small and wiry, mischievously stuck his foot out, causing the largest of the newcomers to stumble. He lunged forward, the rest of grabbed our chairs to defend ourselves when, from nowhere it seemed, a posse of delightful cabaret girls sprang between us, each one leading us by the arm to an exit. It was as well. The big man happened to be the heavyweight wrestling champion of Egypt.

That case became more dramatic thereafter, reaching its denouement when the dahabieh was raided on 12th August 1942 and Eppler and Sandy were arrested. The whole episode has been written up in Sansom's book, 'I Spied Spies', described in David Mure's book 'Practise to Deceive' and made into the film 'Foxhole in Cairo' (two versions). Anwar Sadat described his own part as repairer of the W/T set in his book 'In Search of Identity' and in greater detail in his earlier book 'Unknown Pages', which was withdrawn when he became President of Egypt. Ken Follett fictionalised the affair in 'The Key to Rebecca' and it featured again in the remarkable novel 'The English Patient' by Michael Ondaatjie.

263 FSS

Tom Cordery

I arrived in Tripoli just two hours after the town had fallen on the 23rd January 1943 and in the intense turmoil I was given to Sgt. Banning to go on road checks. He, like myself, was a London TA member (with Westminster Dragoons) – and we sallied forth to check transport for German/Italian deserters, infiltrators trying to move back east, behind our lines, some perhaps contemplating sabotage. Work was of a haphazard nature, constant raids from information received from names on the 'white list', or just looking for individuals who were on the 'black list'. You became inured to seeing women pinching their children to make them cry, so as to draw the crowd's attention, and hopefully distract our attention. Our work was almost entirely civilian, the Div. Sections looked after the military side of things. With the visit of Winston Churchill I was given three 'Jocks' from the Argyles and we had to clear a street near where he would be travelling. This resulted in an altercation with the Spanish Consul, and in another case we allowed an expectant 'Mum' to stay put, rather than to turf her out onto the streets.

On 15–17th February, Montgomery held a big conference for 'Tactical Discussions' at the Albergo del Meharin – a hotel on the eastern seafront. An impressive gathering of generals and staff officers from all departments in the desert advance, plus a contingent from 1st Army in Tunisia, plus some Americans from the same area. This was topped up with a party from the UK, including the CIGS, Sir Alan Brooke. I was fortunate in being stationed on the inside of the door of the main conference room and was able to hear all the problems experienced on the advance, and how they were surmounted.

Another day we had a small conference in our office and I was on the front door complete with tommy gun, and revolver. The only member I remember was the Brigadier i/c 86 Area (Tripoli) but a filthy looking Arab attempted to come in and when I stopped him he replied in perfect Oxford English that he was expected. I met him again some months later when I was in the Cairo Section. He spent his time wandering back and forth through the enemy lines up on the Tunisia/Libya border – we were to meet many more of his kind later when we were at Zouara. There was no set pattern of work in Tripoli – things just cropped up and were acted upon – raids on houses were the most common of the work load.

Came the time to move and we went forward to Zouara, almost the last habitation in Libya. We were billeted in a bank in the main square. The town had a thriving market which to us was a great boon, as many of the traders seemed to come backward and forward from Tunisia as though there was no war. Information from some of these chaps was most useful – although one had to bear in mind, if they could take information in one direction, they could do so in the opposite.

Time to move on again – this time to Zarzis just over the border into Tunisia, near the island of Gerba, the home of the Lotus Eaters of old. Zarzis had a large Spahis Barracks and we found a billet in some houses which were formerly quarters for the Spahis Officers. Only there a short while, when we heard of the projected attack at Wadi Mareth – so we moved up to a mile or two behind the line on the coastal sector. Actually behind 50 Div, who after the battle had commenced, were replaced by the 51st (Highland Div). The line broke and we pushed through toward Gabes, where we found a house in a square in the centre of the town. Again the usual work – contacting Frenchmen and women who were sympathetic and chasing those who were pro-Axis, which included Arabs. Also a wonderful day, when a mass police, security, from Army, Navy, almost anybody who wanted to joined in – searching houses for home brewed 'hooch'. This had been causing a lot of harm to the troops, and had even resulted in deaths. They made whisky in a day but cherry brandy took a couple of days. Anything we found we smashed the bottles and let it run in the gutters, put a match to it and it burned better than petrol.

I went to the Post Office in Gabes (aiming to buy a set of stamps) and discovered it had been used by the Germans as a signal office. Poking around I came across a detailed list of the forces facing the 8th Army – basically a ration strength.

264 FSS

John Mitchell

Syria was one of the French overseas territories which on the collapse of France in 1940 opted for Vichy. In May 1941 permission was given for the Luftwaffe to use Syrian airfields and for war material in Syria to be transported to Iraq to assist the Rashid Ali revolt in progress in that country. In consequence Middle East Command under General Wavell was forced to act and in June 1941 a mixed force of British, Australian and Indian units together with the Free French Force in the Middle East marched into Syria from Palestine and Syria passed into Allied occupation.

264 FSS was posted to Damascus. Damascus, the capital of Syria, is said to be the oldest city in the world in continuous occupation. It was occupied by the Greeks and the Romans and in medieval times was the seat of the Islamic Omayyad dynasty.

When I arrived in Damascus 264 Section operated mainly in the city and the immediate area, but also had one detachment at Abou Chamat, an old caravanserai about 40 miles east of Damascus, on the desert road to Baghdad, Iraq.

Abou Chamat was the Syrian frontier post and housed, in addition to the FS NCO in residence, officials of the Syrian Surete and Customs and a section of the Gendarmerie.

Living conditions at Abou Chamat were somewhat primitive, the FS

NCO feeding himself or as most of us did sharing meals with the Surete and Customs officials.

The rest of the Section were employed in the city liaising with the French and Syrian security organistions and on general Civil and Military Security duties.

Most senior NCOs were therefore engaged on Civil Security duties in plain-clothes. Attached to the Section, in addition to an Army-employed uniformed interpreter, were four plain-clothes Syrian policemen known as the Brigade de Recherche-Hisham Assali, Fozi Baghdadi, Moheiddine Wasfi and Khaled Idilbi. These were under the directions of the senior or senior NCOs, at the time of my arrival with the Section these were Sgt. Thornton and Cpl. Hacknan. Later, in 1945, I myself directed the Brigade de Recherche.

The Brigade always appeared a well chosen team. Hicham Assali was a member of a well known and important Damascene family heavily involved in politics from whom we regularly obtained useful information on the activities and aspirations of the many political parties and factions and the semi-secret political clubs which abounded in the city. This information was often known to us before being known or reported to the British Security Mission in Syria. Fozi Baghdadi must have known the whereabout and activities of every criminal in Damascus and looked like one himself. Moheiddine Wasfi was a veritable terrier who given an enquiry, some were tough and some nebulous, never seemed to stop or rest until stopped or it was terminated. Khaled Idilbi, the only one of them who spoke English was a first-class policeman in all respects and left us prematurely to become one of the first Inspectors in an all-Syrian Surete.

Khaled spoke English, Hicham and Moheiddine French, Fozi only Arabic. We conversed with them in French and our limited Arabic.

In addition to supplying us with the political information we wanted and assisting in enquiries from BSM or the DSO in Beirut etc., one or more of the Brigade accompanied us when operating road blocks and checking identities etc. in hotels, pensions, cabarets even brothels.

Kuneitra is a major town astride the main road between Syria and Palestine and the main duties of the detachment were to liaise with the Palestine Police at their frontier post at Rosh Pinna and the Syrian Surete at Jisr Banat Yacoub. We were also interested in information about the illegal sale of Arab land to the Jewish Agency – Arab families often owned land on both sides of the frontier – and the purchase and smuggling of arms into Palestine for use by the Haganah and the more extreme Jewish groups of fighters.

Baalbeck is the chief town in the Bekaa valley, which is situated between the Lebanon and the Anti-Lebanon mountains. Military security was the main duty as it was here that the larger British military formations were located and the French main air base was at Rayak, a dozen or so miles south.

I was sent to Deraa. In addition to its importance as a frontier town Deraa was the chief administrative centre of the province of the Hauran. On top of his frontier controls the FS NCO in charge was expected to liaise with the local Syrian authorities – the Mohafez or Governor, the Chiefs of Police and Customs, the Commandant of Gendarmerie, and the Mokhtars or headmen of the surrounding villages. There was also a French Political Officer in Deraa who was as suspicious of the FS personnel as were most of the French Administration elsewhere in Syria. The political officers' office was ahout 100 yards away and on the opposite side of the road from the FS office. His obsession with us was so well known that some of the headmen of the outlying villages when they came into town would make a point of crossing straight over the road and into our office after making the customary courtesy visit to the Political Officer, as the local representative of the mandate power.

The end of the war in Europe was in sight and frontier control became less important. At the same time there was increased activity on the part of the Jewish Agency and the Jewish underground movement so that eyes on the frontier between Syria and Palestine had to be reinforced. The permanent detachment at Abou Chamat was withdrawn and only snap checks made there. A new outpost manned from the Kuneitra detachment was located at Fiq, a miserable village in the Golan Heights, overlooking however the Sea of Galilee and the important Jewish colony of Ein Gev. The Golan Heights and the tracks down to the lake were a regular smugglers' paradise – we were concerned with arms, bought or stolen being smuggled into Palestine.

Further new detachments were set up in the city of Homs, between Damascus and Aleppo, a hive of political activity, and in Rayak in the Bekaa valley, which house the main French air base and many French military units.

In May 1945 the Syrians rose against the French. Many were killed and atrocities committed on both sides. The French let loose a Senegalese battalion in the city and almost lost total control of them. The British were forced to intervene and FS NCOs led units of an Indian Division into Damascus. During and after the fighting FS NCOs helped evacuate French civilians and military dependents and recovered much looted property. We often came under fire and were occasionally holed up for several hours at a time – none were fortunately injured.

As a result all French units moved into Lebanon and Syria became fully independent in 1946. The British left the country and 264 Field Security Section was disbanded.

In any operation like Field Security work there are always occasions, incidents and problems which came readily to mind and the following are a few I can always bring to mind.

A fairly common request from other security organistions was for information regarding the movements or whereabouts of individuals of interest to them. This was before the days of surveillance with the aid of cameras. Fortunately Syrians loved having photographs taken of themselves as individuals, families, groups or clubs and photographic studios invariably kept their proof copies. So such an enquiry usually involved a visit by an FS NCO and a member of the Brigade de Recherche who knew the individual involved or his family or acquaintances to all the Damascus photographers and photographic studies and hours of looking through piles of phoographic proofs. We were lucky a surprising number of times.

Searching for or asking about individuals by name was often a challenge because a man would be known to his neighbours or friends by his given name to which would be added that of his father or of his first-born son. This was essential for ready identification given the limited number of popular first names. The man's family or tribal name was rarely used except for official purposes. The same man could thus be commonly known as Mohamed abu Ibrahim (Mohamed father of Ibrahim) or Mohamed ibn Ismail (Mohamed son of Ismail).

In 1944 the Section was warned one day to be ready on a certain date for an important security operation, the guarding of a VIP. We were all to be armed and in uniform. All we were told of the VIP was that he would be arriving in Damascus, staying overnight at the

British Minister's residence, and leaving the next day by the Nairn trans-desert coach service to Baghdad, and we would recognise him when we saw him. FS would supply the prowler guard in the garden of the residence, guards inside the residence and an armed NCO to sit behind the VIP in the coach to Baghdad. In Damascus the Section usually operated in plain clothes. One NCO had not worn his uniform for years, I think, and had some difficulty in finding all of it. Wearing it he was almost unrecognisable to his colleagues. The VIP turned out to be Ivan Maisky, the Soviet Ambassador in London going on to the USSR through Persia after the Cairo Conference. The writer spent half the night sitting on a chair outside Mr Maisky's bedroom, which was across the corridor from Mrs Maisky's bedroom.

When the Anglo-American Plan for the Partition of Palestine was announed in 1945 FS NCOs near the frontier with Palestine were asked to visit Arab and Jewish communities affected by the Plan and attempt to gauge the local reaction. One FS NCO is said to have visited a Jewish colony called Naharia which if the Plan were accepted would become a part of the Arab State.

Naharia was one of the earliest colonies in Palestine, established by German Jews during the first wave of immigrants at the turn of the century. The response he is alleged to have got from the elders as to their reaction to becoming a part of Arab Palestine was to the effect that "Arab or Jewish, Naharia would always remain German".

268/417 FSS
Oswald Ormsby's Magnificent Section

Hugh Dovey

The five FS Sections (264–268) which sailed from the Clyde in April 1941 all assumed that they were headed for the Middle East. But in Lt O.E. Ormsby's 268 Section we could not have even guessed that we would end up in Beirut: the campaign to occupy Syria and the Lebanon and put an end to the German activities that the Vichy authorities had allowed was not launched until June. It ended on 15 July: we were in Beirut the next day.

We quickly found we had several masters. Formally we were centrolled by Army (soon 9th Army) HQ. There was also the British Security Mission set up to assist the Free French who took over the Mandate and were responsible for internal security. Finally came the Defence Security Office (DSO) which represented the masterminds of Security Intelligence Middle East (SIME).

Our first big job was the Vichy convoys. Vichy troops and officials who did not join the Free French were repatriated in August and September. Troops were allowed to take their sidearms, civilians their personal possessions: foodstuffs, gold, military equipment, maps, codes and photos were barred. So with help from two Australian

Sections we spent long hot days at the port inspecting crates and suitcases – and finding little.

When the ships arrived from France they were under tight control, to prevent "enemy agents or other undesirable elements entering or leaving". Our job was also to seal the wireless and locate (unspecified) "devices". Two NCO's stayed on board. It was no chore: some crew member was usually keen to talk to the British, share a bottle and provide information from France. One day one of us was surprised to see among some civilians boarding the ship one who was clearly Sgt Williams. He had come out from the UK in one of the other Sections, had joined us in Beirut without explanation and then mysteriously vanished. Now in mufti he was obviously on his way to other duties in France. (He survived.)

After the convoys routine jobs took over. The security of Service and important civilian installations had to be watched. High on the list was Beirut port where we had a detachment from the start. Better fences and a new pass system were soon introduced. Later when we had a detachment at Tyre we kept an eye on the security controls at the Palestine border. When the Army built a railway line from Haifa to Tripoli we regularly checked the identity of passengers.

We were always interested in reports of Axis sympathisers. The Abwehr, headed by Roser and the notorious Paula Koch, had been busy in the country before we moved in, getting a ready hearing from nationalists who wanted independence from the French or were hostile to the British because we had put down the Arab revolt in Palestine. Some of the more dangerous Abwehr contacts were known to the Army and the French and were soon interned at Miye Miye, some thirty miles south of Beirut. In October 1941 we increased our nearby Sidon detachment by an NCO who would work at the camp. In 1942 the job went to Corporal Maurice Oldfield. His reports from the camp earned him a commission in the DSO. A Lt Colonel in SIME at the end of the war, he transferred to MI6 and eventually became its Chief. Undoubtedly our most famous Old Boy.

A lady who aroused doubts at an early stage was Mme Holstein ('Honky-Tonk'), a delectable Vietnamese who was suspiciously eager to display all her charms to Allied officers. But our report concluded that she was only after their virility and their money – not their military information.

It was not just information gathering – there were sometimes surveillance and arrests as well. The Section records for November 1942 show, without details unfortunately, the arrests of "a German in the YMCA" and of Volontsevech, after he had been shadowed for two days. In the early hours one day in May 1943 we joined the Sûreté in an operation, organised by the DSO, to arrest a number of people in Beirut. There was no need for us to know just what it was about and our orders were of the "make sure nobody gets away by the back door" variety. It was in fact the round up of the Beirut end of an espionage network controlled from Istanbul by Mohammed Uf under cover of his news agency, the Orient Express.

We always kept tabs on civilian morale in general and the degree of support for the Allied cause. Both rose when we were doing well in the war and the supply of foodstuffs was assured – and vice-versa.

In 1943 elections in Syria and the Lebanon produced nationalist majorities and strong pressure on the French to end the Mandate. In the Lebanon the French overreacted and arrested leading nationalists. The country was close to rebellion and 9th Army looked to its FS Sections for first-hand information on the crisis – especially from us at the heart of the storm in Beirut. We instituted daily patrols and reports. Not until the French released the nationalists did things get back to normal.

There was soon talk of a possible move, to the UK perhaps, for the Second Front. In March 1944 28 Section took over, we travelled down to the Depot at Helwan and 268 was formally disbanded.

It had not been all work in the Lebanon. For our leisure hours there had been good bathing in the long summer, sightseeing, restaurants, cinemas, private hospitality, womenfolk. In the billet, bridge and a well-stocked bar, we dined together at Christmas and on 16 July – anniversary of our arrival in Beirut. Ours was an unforced solidarity – the result of Captain Ormsby's light but decisive touch. When he left at the end of 1942, George Harnden, previously the CSM took over and knew just how to maintain the tradition. Old hands and newcomers were sure it was a good Section to serve in.

Did we do a good job? In his autobiography Geoffrey Household wrote:

> "In the years between 1941 and 1944 the Field Security Wing in the Middle East preserved a standard of common sense, discretion, tirelessness and gaiety which made me proud to

belong to it . . . partly it was due to the extraordinary quality of the Sections which arrived from England in the spring of 1941."

Household was in a position to judge. He was an experienced FSO when he came up to Beirut to take charge of security for the repatriation of the Vichy troops. 268 was the first of the newly arrived Sections that he met. And writing about us in that same book he christened us "Oswald Ormsby's Magnificent Section".

Our dismay when 268 was disbanded was shortlived. The Deputy Commandant of Field Security in the Middle East was now Major Ormsby. He was not going to see his old Section dispersed: all but one of us in the newly-formed 417 Section had served with 268 in Beirut. It was with a complete sense of continuity that we landed at Taranto on 4 May, then made our way to Torino on the Sangro, where we came under 5th Corps. That was on 11 May: General Alexander launched his offensive that night.

We had no briefing but we soon found that things were quite different in Italy: no BSM, no DSO, no Sûreté', only the Army. Records now available show what we learned rather piecemeal about the job the Allies had on their hands: "The enemy has made use of agents on a scale far exceeding that experienced in any other theatre either in this war or the last". Against this formidable onslaught there were strong defences: Interrogation Points where agents posing as refugees might be spotted; a special section at CSDIC (Combined Services Detailed Interrogation Centre) for interrogating captured agents; Special Counter-Intelligence Units to penetrate enemy intelligence services, using "turned-round" captured agents, for instance; SIM, the Italian Secret Service, now giving the Allies the benefit of its local knowledge and experience. But the FS Sections and the CIC Detachments, their American counterparts, were the 'backbone' of the system.

Our main source of information about the enemy was the CI Bulletins that came to us regularly from HQ. They gave us the names, descriptions and particulars of enemy agents to be added to the 'CI List' (a card index in practice) and of the captured ones whose names could be deleted. Their cards went in the 'Dead Box': it grew at an impressive rate.

Some agents were left behind by the retreating Germans. But the majority who came across the lines, landed on the coast or were parachuted down had to travel to reach their targets and to return to base. So control of movement was essential. The nearer the front line the tighter the restrictions on travel without a Military Government pass. One was needed to leave or enter the Army area. When we arrived the control line was along the Sangro and we ran a series of roadblocks at the river crossings manned by Carabinieri or Italian troops. Civilians travelling without the necessary pass or any means of identity – an offence in itself – were brought into our HQ or detachment offices for questioning and checking against the CI List. We soon learned the drill. We had full powers of arrest and sent the 'Category A', travel without a pass or means of identity, offenders off to jail to appear before a Military Government tribunal.

We were on the look-out for bigger fish: prominent fascists and other 'undesirable elements'. We sent these 'Category B' detainees off to internment camps in the south. There was a special drill for captured agents.

As the Germans withdrew we moved into Sulmona. Many Allied PoW who had escaped from the nearby camp at the Italian Armistice had been sheltered and fed by local people at great risk to themselves if the Germans found them out. Nadia Bufarina, a prostitute, had been notorious for betraying helpers and collecting the reward the Germans paid for the information. She had fled with them. In Aquila, our next stop, the German Town Major, whose office we took over, had obligingly left us the address of the local Wehrmachts-Bordell (Army brothel). Nadia was not known there but the Madam thought that Gina, who was off duty when we called, might have been in Sulmona and offered to send her to us. At the office next day Gina was anxious to explain why she was a prostitute: the Germans had forced her into it. To demonstrate her good character she displayed her left breast on which were inscribed the name and address of her fiancé. What interested us even more was that she did know Nadia and would bring her to us. She did so two days later. With a minimum of formalities we sent Nadia on her way to internment. Gina got our profuse thanks. Our satisfaction was shortlived: we heard from No. 1 District, into whose hands Nadia had passed, that she was being released: under the Germans, concealing the whereabouts of an escapee was an offence, and she could not be punished for keeping on the right side of the law. Our protests were vain. They might have succeeded had we known that No. 2 District had just confirmed the internment of two men arrested in Chieti by 427 Section for doing just what Nadia did.

With the Control Line running through Ascoli Piceno we moved up there with our newly acquired detachment of Italian troops, commanded by the voluble Tenente Casalino. Our road blocks were busy: 4800 civilians checked in one fortnight in July led to 23 Category A arrests. We had five Category B in the same period, small beer compared with the totals of 398 in June, 875 in July and 476 in August reported by HQ. The Allied advance had caught up with a lot of Fascists and the like. More important were the figures for captured agents: 100 in June (75 of them in Rome), 32 in July, 52 in August and 18 in September. One a day was now the average catch since January.

September saw us in Cortona in central Italy. The railway to Arezzo ran through our area as well as some main roads. Passenger train checks and the usual roadblocks resulted in 32 arrests in October. But it was in Cortona that an agent slipped through our hands. We treated him as a normal 'travelling without a pass' case and sent him to jail. While he was there he was listed in a CI Bulletin but his name did not ring a bell. At the end of his sentence he continued his journey northwards: he was returning from a mission. A Section in the forward area picked him up and asked us to confirm that he had been in our jail. HQ tightened the procedure following this case: thereafter no one was released from jail until FS or CIC had checked again. It's an ill wind.

After Cortona it was Arezzo, Army railhead and provincial capital. Our roadblocks there were the busiest we ever had. A new jail had to be opened: the old one was overcrowded. We had good cooperation with the Military Government officers. Together we brought in a new CI List check: they sent us the travel passes they proposed to issue before giving them to the applicants. That gave us the chance to question any whose names raised doubts.

A dozen or so Fascists we sent to internment included Mugnai, the local Party Secretary. He emerged from hiding hoping in vain that his past had been forgotten. Best of the Arezzo arrests were those of two suspect agents. The one on the CI List had vanished in Bari after arriving in a batch of refugees from Albania. Planted by the Germans perhaps. Our very efficient SIM pair tracked him down to his Arezzo home. The other was a Montevarchi girl whose name and address were in an agent's belongings when 427 Section arrested him in

Florence. When we first interrogated her she denied all knowledge of him: it took some patient questioning to make her confess. We passed her on to 427. Ours was a small contribution to good overall figures: 23 captures in October, 55 in November and 48 in December – well over one a day. All 15 trained in one Abwehr course in October were caught: to celebrate they were mustered for a group photograph in Forence jail. There was no dramatic reason for the good figures, simply more roadblocks controlled by FS and the CIC in the forward area. The repeated checks were enough to make some agents lose their nerve and surrender.

HQ had continued to feed us with information about the enemy's methods. One, posing as a doctor, opened a 'surgery' to provide a port of call for others. Some were instructed to get jobs with Allied units and one had in fact worked in a British officers' mess. (Vet all civilian employees was the lesson.) One 'refugee' had a transmitter in his suitcase. Another had been badly briefed: he told the same cover story as an agent already captured and carried a 1000 lire note in the same series as the first one. Father Biondi looked innocent enough when he arrived from the north with letters to the Vatican and his Order in Rome but he had nothing to justify his attempt to travel back and had to confess that he had been on a political espionage mission. Lt Pfannenstiel of the Abwehr dressed himself up as a shepherd, borrowed a flock and took it across to the Allied lines to see things for himself.

Women agents were prominent. Most remarkable was 17-year-old Carla Costa. Her first two missions had earned her the Iron Cross and Mussolini's personal congratulations. She was as tough as her record suggests: captured on her third mission she withstood five days' intensive questioning before owning up. Less spectacular was the woman who worked for an Allied unit in Florence and sent back a number of messages before she was caught. Agents needed a means of identification when they got back to the German lines. Women, HQ reported, had a small piece of cloth with identity particulars written on it. It was sewn in the elastic top or hem of their (Br) knickers (= US panties). We laughed at the joke – and rubbed our hands at the prospect.

In 1945, even with the end of the war in sight, the Germans did not let up: 26 agents captured in January, 32 in February, 60 in March – including the youngest ever, a 13-year-old boy. The Allies' final offensive took us back to the Adriatic, at Pesaro and Ravenna, just when the enemy organisation was collapsing: 41 agents captured in April and a stupendous 210 in May. A lot had retreated to Venice where 47 Section alone rounded up 50 in May and 17 stragglers in June.

Peace brought a rapid change in our preoccupations. We were in some of the first skirmishes of the Cold War, long before that term was in common use. On 20 May the partisans formally stood down and were supposed to hand in the arms the Allies had supplied. The Communist Garibaldi Brigades were very reluctant to do this. There were fears of a Communist coup, especiallv when they surrendered far fewer weapons than they were known to have. This was of particular concern to us in Ravenna: the Garibaldi Brigades had been very strong – and, it must be said, very effective against the Germans – in the area just to the north. We were soon getting reports of hidden arms dumps.

One report, about a farm near Alfonsine, was precise and circumstantial. With some of our Italian troops we raided it on 10 July. Under the hay in a barn were some 50 rifles, some light machine-guns, ammunition, grenades, plastic. But the farmer, Mazzotti, had vanished. He must have seen us coming. We were content to let him think we had forgotten him. Late in September, very early in the day, we took the troops back there. There was still no sign of him until Tenente Casalino thought to turn back the bedclothes on a bed that had seemingly not been slept in, found that the sheets were warm and discovered Mazzotti the Red under the bed.

We put him in jail to await trial. Bologna jail: we had moved from Ravenna earlier in the month. It was our last station. Arms raids were still on the agenda, though no haul was on the Alfonsine scale. Straight political reporting came to the fore, with Communist activity a preoccupation. The Russian Repatriation Commission was secretly stirring things up. Our excellent SIM Detachment did good work for us at this time.

But the Allies were loosening their grip on security. Our ranks were depleted by demobilisation while we were given an ever growing area to cover. By June it stretched from Rovigo in the north to Ancona way down the Adriatic. And in that month 417's records came to an end.

Italy was at least as good as Beirut for creature comforts. A house was always to be found that served as billet and office, with civilian staff to cook and clean. There was usually wine and vermouth and sometimes female company. There was more culture: sightseeing and music. We continued to celebrate Christmas and the 15 July anniversary of our arrival in Beirut.

As Sections went we were not outstanding, perhaps because we arrived late, with no previous training, and our job back on the control line offered fewer opportunities than those of the forward Sections. But we ran our roadblocks efficiently and did well to capture Nadia (even if she did get away with it) and Mazzotti. Best of all we played our part in an outstandingly successful system. Geoffrey Household referred (rather enviously) to Sansom, his FS colleague in Cairo as ". . . one of the few of us who really did justify his existence by occasionally arresting an undoubted spy". (Sansom's Section arrested Eppler and Sandstede, the Abwehr 'Operation Condor' agents in Cairo in July 1942.) Nothing could illustrate the different situation in Italy more strikingly than the two half-monthly reports of 278 Section for March 1945 in which the FSO recorded without flourishes the arrest of three and four agents. Household's final judgment was that FS was at its flowering in the Middle East. In Italy it fruited richly.

265 FSS

Jake Jacobs

The Field Security Section that I joined at Sidi Haneish in December 1940 was commanded by the same Busty Peel who had travelled with me from England a year earlier as a Lance Corporal and had spent a few weeks with me in Suez before being commisioned. Section 265, the first to see action in the Middle East was creating a role for itself as it went along. It worked through GSI (a) (offensive intelligence) and was required to be as mobile as the Western Desert operations themselves. My arrival was happily timed to occur a few days after General Wavell had started his astonishingly successful offensive at Sidi Barrani. We worked in small groups of two and three, sometimes attached to various units. Our task was to obtain, largely from captured documents, as much information as possible concerning enemy dispositions and get it to forward Intelligence HQs with a minimum of delay. After a few days at Mersah Matruh I was told to make my own way west towards Bardia which was expected to fall shortly. From December 18th it had been subjected to the light fire of our mechanised units. I remember digging a slit trench for myself outside Sollum feeling rather sorry for myself because it was Christmas Eve. I contacted my CO next morning having been awakened by the artillery barrage which was wearing down the extremely strong, natural fortress. We were given aerial photos (I have mine still) of the town with a number of military headquarters and civil administration buildings identified. Our task was to get rid of them as soon as possible and gain possesion of as much documentation intact as possible. One difficulty was the absence of any communication system to learn how the battle for the town was faring. In effect we were required to depend on our own visual observation. On this occasion the result was highly successful and amusing. With a less dispirited enemy it would have ended disastrously for me.

On January 3rd 1941, Busty Peel and I with two New Zealand privates had observed the main assault on the town and decided we would slip in by way of a desert track which approached the harbour from a direction which appeared to be unprotected and certainly uninvolved in the fighting. We accordingly piled into our 15cwt truck prepared in a totally irresponsible way to cross our bridges as we came to them. Within a mile of our objective in very broken country we rounded a bend to find an Italian Company beautifully dug in occupying a commanding position on both sides of the road. Instead of opening fire on us – we were sitting ducks with no possible chance of reaching cover – they waved white handkerciefs and indicated their lack of interest in following Mussolini's dictum of preferring to live like lions for a day rather than as lambs for a century. We called their officers down and formally took possesion of their pistols (I kept one as a souvenir until the end of the war: On my way home to be demobilised I reprehensibly sold it in Paris). The occasion became quite emotional with the officers in tears and one of them producing his wallet and the inevitable family pictures. We told them to stack their arms and then proceeded about our more urgent business. Doubtless in due course they joined the ranks of their 130,000 compatriots captured during that brief, successful, though scarcely glorious campaign.

There are some other highlights of this period which stand out in my memory. My baptism of fire took place during the first assault of Tobruk on the 21st January 1941. The whole section was crammed into its 15cwt truck which was approaching an anti-tank ditch which had been breached when a shell burst perhaps 20 yards ahead of us. In our ignorance we stopped and flung ourselves on the nearest piece of ground – all around the truck. A second shell then landed about the same distance behind the truck, and that was all. After some ten minutes we rather sheepishly remounted and drove on. Much later, when I had learned a little more of the arts of war I realised that we had all been 'bracketed'. A third round should have landed right in our midst. I have often wondered since what distracted the Italian gunners who, I later learned, were by far the most effective part of the Italian army.

That same night I scribbled in pencil the following . . . the original survived, though now scarcely legible, the intervening years:

Into this aching heart of mine they'll doubtless blast a hail of lead
Or, less mercifully, perhaps shell shocked or splinter in the head.
And when at last the blood stops flowing from my wound
And slowly I start to stiffen on the ground,
Will there be found an enemy to say
'I killed that man, I'm, happier today'?

Or will another soldier, leaden eye,
Lie sleepless on his bunk, having cried
Himself to wakefulness, softly and hopelessly,
Wishing and thinking of all that now can ever be?

But there will be many, miles away,
Even, perhaps, some close at hand,
Who will be rendered numb then sad
But never really understand
How I died or how I cried
And why I didn't even pray.

And will this war be nearer won
As we meet death or maiming?
Is not the end sent farther on,
That end at which we're aiming?
Or does each addition to that cursed list
But merely harden broken hearts,
Hearts of those whose lips we've kissed?

January 1941,
Western Desert, Africa.

As this period in the desert was the only time I found myself involved in a shooting war on land I may as well dispose of the only other comparable incidents when I felt that the enemy was interested in the destruction of Sergeant Jacobs in particular. There was the occasion – I think it was outside Benghazi but cannot be sure of this – when I found myself walking along the edge of some recently cultivated land bounded by dry-stone walls. There were some infantry around. We were not under fire. For some inexplicable reason, certainly not conscious premonition, I bent double below the level of the wall. As I did so an artillery shell exploded just the other side of the wall. I was blown to the ground but quite unscathed. A comrade on the far side of the wall was less fortunate.

Finally there was an occasion when my discomforture, unlike the occasions just recorded, was continuing rather than of a moment's duration. The day before the attack on Derna at the end of January, the Section members had been briefed on their duties the following day. I was to report to a PoW cage which it was planned to establish and given nothing more than a map reference to permit me to identify it. I was to take delivery there of such captured documents as would have been collected by interrogating officers and others and deliver them to an advance Divisional HQ, for which I also had nothing more than a map reference. My transport was to be a captured Italian SPA lorry which I would have to drive. In the event, most of all to my surprise, it all worked out as planned. I had to follow the line of an escarpment to my right. As now recalled the trip went on forever and its first half was very much in sight and sound of the battle. The whole journey was of course across the open desert with no question of any roads or tracks to follow. It was with a feeling of enormous relief that I finally reached my destination and found someone to take over the contents of the lorry. Whether anyone ever subsequently looked at them is another matter on which I shall – perhaps happily – ever remain in ignorance.

276 FSS

Gerald Cockell

Gerald Cockell's memories of the invasion of Salerno, Italy in 1943 and these are followed by extracts from his diary, written at that time.

276 FSS was the first to enter Salerno on 9th September 1943, in fact I was one of the first three people. The same night a German patrol was also in Salerno but did not see us. I was also one of the first ten people to enter Rome. This may have led to the custom that Field Security was always, 'First in and last out'.

On arriving in Salerno three of us went straight to the Post Office, there I found a telegram on somebody's desk saying the the Italians expected a landing on the 7th September 1943. We also intercepted a telephone call from an Italian colonel in Naples (then in German hands). He phoned from the brothel and asked for information and whether the Germans still held Salerno.

We contacted the Civil Police and went through their records. I was later posted on detachment to Ravello and got rid of the Police Chief there who was disliked by all, and corrupt. It was while on detachment in Ravello (in a beautiful villa where I later went on honeymoon) that the King of Italy and his suite came to stay in one of the few hotels. I was approached by one of his military guards and had him to lunch but was given strict instructions by my own FSO not to enter into any talks or negotiatons.

I still recall my crossing a bridge on the road Salerno-Amalfi quite vividly. It was under constant mortar fire and a jeep and two motorcycles had been hit and caught fire and abandoned on it. There were two holes in the road made by the mortars and I had to cross to the other side to collect a so-called collaborator. I made it at full throttle with my heart in my mouth all to no purpose as the person I went to collect had disappeared.

While we were in Rome my FSO (Capt. Rafter) was receiving information from MI6 that German spies were still sending information back on our own troop movements. The information was intercepted and we knew exactly what they were sending, I was asked to try and locate them. As their reports were all of movements going north, I positioned myself on a bridge over the Tiber, leading to a roundabout where all the traffic had to pass when moving northwards. I soon discovered that the reports by the spies were totally inaccurate as I also monitored the movements of the traffic. The spies were eventually caught, not by us, although they were in the basement overlooking the roundabout. When interrogated they admitted that their reports were made up after they had been to the cinema. They were caught through a German parachutist who had been sent to give them money. He was captured by our troops and interrogated and gave their position away.

We were overwhemed with work in Rome, we had been given addresses of collaborators and Squadristi (first of Mussolini's followers) and told to arrest them.

There was a constant queue outside our office of people who wanted one thing or another . . . e.g. a mother expecting a baby and wanted a pass for the curfew. We kept on telling people to come back as we could not deal with their enquiries. After about 4 to 6 days we finally spoke to a man who simply said he was a spy and wanted to give himself up.

THE DIARY OF GERALD COCKELL

The Invasion of Salerno, Italy

September 1948 – 276 FSS

Feverish preparations were going on, and soon we were moved to a rest camp where we stayed for eight days and having quite an easy time. After that I was attached to another FS Section which was moving along with a Div. Two days later we were in a transit camp, that was to be my last night on African soil. I can well remember thinking about it then and the thought of leaving all this dust and sand pleased me intensely.

Next morning, 3 September, I was stung by a Green Scorpion which must have spent the night with me. Did I curse. However, the sting is overrated and lasts between 12–18 hours.

The 3rd September will stand out as a date. It was then that war was declared. It was also on that day that we landed in the Middle East and also the day we left it. Our next port of call was to be Bizerta, there however, we did not get an opportunity to land much to my regret. We stayed there one night.

A whole fleet or rather armada, was anchored there, it was a most impressive sight, but what was more impressive still was the air raid we had the following night.

Every ship seemed to open up and the sky seemed filled with a mass of red balls darting all over the place but mainly in one direction. Some people on the ship compared it to Crystal Palace exhibitions but there were more fireworks than that.

It was here that I saw a few enemy planes caught in the beam of searchlights. It was uncanny the way they picked the planes and followed them.

The ship we were on was a Landing Ship Tank, and after leaving Bizerta the crew and passengers were informed of their destination. We had however, already been told at the Rest Camp. Life was well organised and meals were run on the cafeteria system. There was little waiting for the bully beef and potatoes which was dished out in our billy cans.

The journey was uneventful but everyone was avid for news we received and rejoiced when we heard Italy had signed an Armistice. That was the 8th September and we wondered what reception we would get the following day. Our first attempt at landing was at 0830 on the 9th. We cast off the pontoon we had been trailing on this occasion and were told to go back. Then for quite a while we seemed to be cruising round the bay. It was the most amazing experience and I felt very much like a news reporter to the BBC. There, a ship has just got the beaches, the troops and transport are coming off. Look! Did you see that? It was a shell, pretty near her too, and there's another. Gosh, I pity the blokes on that ship, wouldn't like to be there. Look over there! A good mile from shore, those amphibious cars are coming across the water, there they go, one after another. I watch them as they go by, they are quite near now and we can discern them easily. A small fleet of them have come out of the bowels of an LST and are making for the beach.

There is activity on the beach now, little men and transports going to and fro and shells falling quite near them. The beach-master comes up in a little motorboat. "Are you for Roger Beach?" he asked. The captain makes some reply and shortly afterwards we were towards the beach again.

Gosh! That was deafening! It was our warships opening up. They are firing quite regularly now. Full speed ahead, we are going to ground the ship on the beach. Everyone by their vehicles please, I'm thirsty, my throat is parched and moreover I've got a lump in it. Mind the bump, we're nearly there, here it comes. Oh! not really what I expected . . . the doors are opening, they are open now. But why isn't anything happening? Nothing is coming out.

Oh but hurry for God's sake, or they'll get our range. I'm thirstier than ever, and I've just noticed someone drinking tea. For the next 10 minutes I'm hunting round the top deck to cadge a cup of tea. At last I'm successful. Just at that moment I hear we are going to make another attempt at landing.

I gulp down a few mouthfuls of hot liquid and burn my throat and lips. We are off and just turning around.

Wham . . . wham . . . wham! Three of them which I don't like a bit. Trent says they fell 10 yards behind the ship. We seem to be out of range now and cruising around. Some ship is laying a smoke screen, looks like she is on fire but we know better. Quite a few other ships seem to be following the example.

Shelling seems to have lulled. We are having another go. Full speed ahead . . . one . . . two . . . three . . . we've grounded. The beach is being ploughed up by all sorts of vehicles. REs are laying down wire netting. There is a traffic jam. It's now about 1630 and at last my turn, I start up the bike with the first kick. I'm out on to the sand and passing numerous vehicles lined up one behind another; over a little bridge which crosses a dyke and away. My instructions were to rejoin the FSO at the assembly point or Div. HQ. I enquire at the first corner. No one knows for certain but believe it to be to the left. I consult my map and go. I'm halfway down a lane now. But isn't it quiet! Too quiet, I turn back as I reach the crossroads, a captain stops me and asks me the situation at the bottom. He says there was bloody street fighting and enemy tanks. A bit shaken I take a left fork and after a bit hear shells whizzing over me so I take cover in a trench. A signals officer joins me and we have a smoke and a chat and I confirm my position on the map and that of Div. HQ. There seems to be a lull so I take a chance and dash down the road on the bike and make it to Div. HQ. Here is a hive of activity, people digging trenches. Big guns take up positions, camouflage nets being put up. FSO arrives and we stay for the night. Shells burst as we dig our trench, guns open up, mosquitos abide in thousands. I make my bed under a lorry in close proximity to the trench, pull the mosquito net over my head and I'm asleep. Once during the night I get up for shelter in the trench. Next morning I wake up early and cook breakfast on a primus. Bacon and tea. We hang around all morning. At lunchtime I'm in a farmhouse eating omelette and macaroni. Late in the afternoon we set off for Salerno. The situation there is still not clear but we risk it. We arrive at 2000 hours and immediately set to work. Contact civil police, secure telephone exchange and post office. Tired out we get to sleep at 0200 hours.

Next morning we find a billet in the town, make ourselves comfortable and set up office. What a change! Wine in abundance, white table cloth. Three women look after us. Plates are changed at every course. Luxury, yes, but all the same the war is near and shells whine over our heads, burst among the buildings and is a constant reminder that all is not well. We work until late at night but in spite of the terrific din, sleep tight on comfortable beds.

More troops come into the town now and it gives us a feeling of confidence. We were the only British people to sleep in Salerno the first night (10 Sept). For the next few days salvoes of shells come whining over our heads at intervals. Now, however, (21 Sept) the shells have been replaced by mortar bombs, they don't whine as much, in fact you can't hear them coming. When they burst they make as much noise but less damage. They are very unpleasant and frightening.

Days slip by at a terrific speed and the work is exceedingly interesting and exciting, sometimes. The other day I was sheltering in a house which was hit six times by mortar bombs.

Only yesterday my heart was in my mouth when I had to cross a bridge which was under mortar and machine gun fire. I got through both times but felt terribly scared.

Refugees are streaming in and have been for some time now. It's a terrible and pitiful sight. We get very many into the office and all our cigarettes seem to disappear. They have no food either and a few haven't eaten for several days. Others have walked here all the way from Rome.

Later.

Our days of excitement are over now, the Div. Sections have left us. The battle has moved on, all is quiet. There were many things which happened which should also have been related. Other members of the Sections had their own exciting moments.

All that is left now is the babbling of voices of the refugees. They want this, they want that. Where can they go for this, what happens in case of so and so. Curfew has naturally been imposed since the first hour and now an expectant mother wants a permit to allow her son to call for the doctor. You would really be surprised to know all the things they come to us for.

I took a trip into the country . . . it was simply beautiful.

With 276 FSS in Italy

Tom Bewick

276 FSS took part in the invasion of Italy in 1943 at Salerno. Shortly after the landings I was sent to look for a contact who had reported that a British Commando was hiding in a cave with a sprained ankle and would I take him to the Special Service Brigade which, at that time was in the south of Salerno. This comprised No. 2 Commando and 40 Commando who were resting after their battle around Vietri with the retreating Germans. I eventually found the soldier and escorted him to the rest area.

I was also informed to return to the Vietri area and make contact with a pair of Italians who turned out to be father and son, respectively named Luigi and Antonio, who were working for the American OSS (Office of Strategic Services) and cultivate friendship with them.

After doing civil and military security duties in the Salerno are I was advised by the FSO, Capt. Rafter, that I would be going on a special mission with the two OSS agents to locate a Col. Zaniboni who, in the 1920s whilst serving in the crack Italian regiment, the Bersaglieri, had attempted to assasinate Mussolini and been banished to the Isle of Ponza and was now required by the British and Americans for the Psychological Warfare Branch. The OSS men and I found the colonel and his daughter in a hotel on the isle of Capri. We all returned to

Salerno and I left with the OSS men. Capt. Rafter provided me with documents to draw petrol and arrangements for hotels in Bari. Miss Zanaboni would be staying with friends at a villa in Salerno until her father reurned.

I can never forget our return journey as in every town and small village, crowds were on the streets waving, cheering and screaming for Colonel Zamboni to make a speech. He was a most distinguished figure of a man, six feet two inches in height with iron grey hair and the crowds wanting to shake his hand. No wonder the PWB wanted him.

With 'S' Force in Italy

276 FSS was chosen as one of the unit to form 'S' Force to go in and occupy Rome. The special force was commanded by Col. Smith of the US Army who gathered all the various units together at a special meeting in Naples to tell us that he had the honour of commanding the Force to occupy the first enemy capital in Europe. This force of American, British and Italian intelligence units entered Rome in the early hours of 4th June 1944 and had passed through a blazing Cisterna which had been the scene of a bitter battle between American Rangers (Commandos) and German forces. We then proceeded by the Via Appia and the Alban Hills to Rome. The section established an office in the Via Sicilia which is off the Via Veneto and the building was shared by various intelligence units; there were both American and British. I was told I would be working with a Capt. Soames on special intelligence duties in Rome. The captain turned out to be Capt. Christopher Soames of one of the Guards regiments and was engaged to Mary Churchill, daughter of the Prime Minister.

We had a list of 5 or 6 Italian fascists who had been trained in sabotage activities in northern Italy and were now believed to be in hiding in Rome. We made a number of enquiries and traced and arrested the agents and lodged them in the Security Wing of Regina Celli Prison in Rome. After working together for 5 or 6 weeks, Capt. Soames and I said cheerio to each other and wished each other good luck with a farewell drink.

The section had taken up residence in Vialle Rossani which was in the Parioli quarter of Rome and the villa had previously been occupied by the staff of the notorious Col. Kappler. The SS Colonel was responsible for the massacre of many in the Ardentine Caves on the outskirts of Rome and one of the victims was the Air Marshal of the Italian Air Force.

I was then sent to screen refugees at the centre at Cine Citta as we had been receiving reports that certain factions from Jugoslavia were causing trouble. This was proved to be the case as a former inmate at the centre who had served as a Chetnik with General Mihailovich recognised three Matashi generals who had helped the Germans in Croatia and were now trying to pass themselves off as refugees. After screening them I had their belongings searched and found enough evidence to have them arrested and handed over to CSDIC . . . Combined Services Detailed Interrogation Centre.

276 FSS
Italy (Salerno–Rome) 1943–1945

John Martin

When 276 FSS landed on the Salerno beaches it was hoped we would be in Naples within 3–4 days. However Kesselring's prompt deployment of his forces delayed this and the occupation of Naples was not until some weeks later, and there was a change to the section's future movements. We were left to sort out a few problems in the Salerno area and it was soon realised that the 'too hard basket' problems were foisted on to the Section to clear up. Consequently I was ordered to look after the security at the small port of Salerno, where a maximum of four ships could discharge cargo and smaller Italian coastal boats called for shelter, one of these was still carrying an Italian gun mounted on its foredeck, which after some discussion, I removed from the vessel and brought it back to the office where it stayed for some time until we handed it back to the 'Carabiniere'. The CSM was muttering about cluttering up the office with usesless equipment and my good intentions of disarming a former enemy were not appreciated. I also had to deal with the deserters from various merchant ships who having been given shore leave after crew lists were checked either became very drunk or ended up in an 'off limits' brothel and they missed their sailing time. One even managed to stay in an army transit camp for some days before he was picked up.

The Navy and Military Police washed their hands of this problem so it was left to the section to clear up, and some summary justice was meted out and after a few days in the local lock up, the deserters were quite happy to get on the next boat available and word must have gone around to make sure not to 'miss their boats' in Salerno.

Another problem I had to sort out had a minor controversial but for me a satisfactory sequel some months later. Shortly after the section had established its office at the Villa Lamberti, a young French boy wandered into the office. He was age 10–12 years and after questioning it was revealed that he had made his way through France into Italy, through German and Allied lines, with the object of joining his parents who were living in Algiers. He had tired of living with his grandparents in German- occupied France. I was told to find a boat and send him on his way. This meant daily checks with the RN and MN offices, neither were very helpful, their hands being rather full at the time to worry about a French stray, so for a few weeks he stayed with the section. One evening I was walking along the beach at Salerno with him. By chance an American LST was on the beach discharging various stores, among which there must have been a consignment of whiskey. I made a casual remark to a crew member on the beach as to their activities. He offered me a pull from a whiskey bottle he had broached from the cargo and this established the necessary rapport, and I learnt that the LST on completion of its discharge of cargo would be returning to Algiers. I took the French lad aside and in my best French explained what I had learnt. The LST sailed later that day and the next day no one really noticed his absence and I assumed his Gallic charm with the Americans ensured he was on his way. With the daily changes and other problems this incident passed into the unit files. I had almost forgotten it until April 1945 when with Capt Rafter FSO I was attending a course at the IC Depot Castellamare on Austria, Austrian Police and Intelligence organisations. At the time there was talk of the section moving to Austria. One of the lectures given by a Major whose pet subject was Canalisation and it's importance when the war was over. To demonstrate how it did not operate according to his book of rules he quoted an incident which had occurred in Algiers. A young French boy had arrived on an Allied vessel with no papers of any kind and was able to move through the lines in Italy with no apparent security clearance after he came into Allied territory, everyone must have been at fault according to the Major. Capt Rafter gave me an unobtrusive 'thumbs up'. We at least knew the outcome of our Salerno problem child. We both kept quiet as the lecturer was now in full flow and would have no doubt been a little peeved if we had elaborated on his story. I hope young 'Jacque' survived the war and now enjoys his senior years in France.

Steve

The late Malcolm Muggeridge claimed that many members of the Corps were drawn from the ranks of schoolmasters, journalists, unfrocked clergymen etc. I am not quite sure which category Steve would have fitted.

I met Steve for the first time travelling on a train from the West Country to join the Corps at Winchester in 1941, where we were greeted with some dismay by Cpl. Mostert, as Steve was not quite up to Brigade of Guards standards. We were in the same squad at Winchester and after completion of training we went our various ways and did not meet again until 1944 when he showed up in Rome to join 276 FSS. Although Steve was from a South American country he did not carry the 'BLAV' flashes. He claimed he was conspicuous enough without this display. He was fluent in Spanish, Italian and French. He claimed he had taken part in the the Gran Chaco war, Bolivia/Paraguay 1932/35. English was his weakest language and by any military standards he was scruffy and a bender of military regulations. He was a great 'womaniser' but did tell me that he did not join us in our evening drinks in the mess as he got into enough trouble in a sober state, and feared the results if he drank too much. Somewhere on the way he had parachuted into Southern Italy ending up in a small Italian community where his principal activity seemed to have been with the local girls. On his arrival in Rome he announced that he wanted to marrry a girl from the South. At this time the Army frowned upon local marriages but this did not deter Steve and he was duly wed in Rome and the girl returned to her home.

As far as his security duties he was a disaster. He had difficulty in getting things down on paper and even on the most routine of enquiries Steve always seemed to become involved with a woman however remotely concerned she was with the enquiry. The CSM was heard to declare "If Steve was sent to the remotest Saharan outpost he would get mixed up with a woman". Eventually he was returned to the

Depot at Castelamare his time having run out with 276 FSS. He had however acquired a flashy Fiat car and a 'blonde' Roman lady during his stay with the Section. His departure however did not entirely sever his links with us. One Sunday afternoon not long after his departure as Duty Officer I was summoned to answer a call at the front gate of our billet and there I was accosted by an irate Italian who demanded to know the whereabouts of Steve, "Where was he, he has stolen my car, he has taken my wife" and folornly added also, "la biancheria" (bed linen) which was the final insult. To the best of my knowledge I told him Steve was with the army well north of Rome and was unable to give him any further news of Steve. The disgruntled cuckold Italian then went on his way. Some weeks later there was a similar summons to the front gate. This time it was Steve himself, still scruffy but with his usual aplomb. He was driving the flashy car and was accompanied by the Roman 'blonde' who by this time had acquired a British Army ATS uniform. Steve was once again on his way back to the depot and was in dire need of petrol. He was able to produce a 'work ticket' for his car, however spurious it was and he was a little wary of presenting it at the army petrol point. By then our section driver 'Bert' was called in and with his help a tank full on the Section truck, some rubber tubing, petrol siphoned a discreet distance from the petrol point Steve proceeded on his way quite happily. Some time later in early 1946 after I had left the army I was working in Naples with a UN Mission and called at the depot where Steve was awaiting discharge. By then he had acquired a marked impairment of his legs and moved with a pronounced limp which was confounding all medical examinations but Steve was still enjoying his liaison with the lady from Rome. I later learnt from a friend who was still with the FS Section operating in the Naples area that Steve had nothing wrong with his legs and indeed had shown him that he was still very agile but Steve was a little apprehensive of the Military Police enquiring into some of his activities. From a Military Policeman acquaintance in Naples I learnt they were never quite able to catch up with Steve. He always appeared to be one jump ahead and they were never able to nail any firm charges against him. This was the last I heard of Steve. I doubt if Steve would qualify for a meritorious mention, but he was a friendly likeable rogue and could probably be listed as a Corps 'Character' who managed to survive his military career in the Corps which no doubt had many other similar misfits.

Rosetta I—li (Rome/Naples 1944–46)

Her name showed up in some German documents which came into the office and was passed to me to deal with. After checking around I was able to find her last known address and eventually track her down I was able to interview her at a flat which she was sharing with her sister. Rosetta was a a young bright and clever Neapolitan, well-groomed, very feminine and attractive. Originally from Naples, her family had moved to avoid the Allied bombing to shelter at a small mountain community at Roccamonfina in the Arunci mountains well north of Naples. The village was later to become within the area which formed part of the German defence line as the Allies advanced on Cassino and what had been a refuge for the family soon became an important point on the line and the German army moved in bringing some harassment for the locals and the conscription of young Italian males had already begun. Rosetta was a student at the Naples University, in the language faculty and already proficient in German and working hard on her English. When the Germans became aware of this she was offered a job as interpreter with a German unit engaged in property and local supplies requisition, which had caused some problems with the local population. Rosetta thus found herself in some quandary, on the one hand she might help ease some of the problems but at the same time be regarded as a collaborator. In addition her two brothers were likely to be rounded up by the Germans. As she pointed out to me, it was a difficult decision for a young nineteen year old to handle and something the English had not to contend with in their country and would not really understand how she felt. She decided to take the job. Her brothers escaped the clutches of the Germans and she found her duties rather innocuous and with her native Neapolitan cunning avoided any major disasters. She did however form a friendship with a young German officer in the course of her duties. She did stress at all times the Germans treated her with great respect and never had any fears of harassment. When the Allies broke the German lines and the retreat was on she was able to get away as far as the war was concerned and she was most anxious to get back to the family still at Roccamonfina. She had no great political feelings and to me she no longer was a security threat. I made my report to the FSO on these lines which he agreed with. He did however add "Keep an eye on her as long as she remains in

Rome". This was a pleasant assignment and from time to time I was invited to visit the sisters and met other members of the family and their friends. We discussed politics. Later on they could never quite understand why Churchill was given the push. They all had great interest in England, the Royal Family and the men were well informed on Soccer and knew all the prominent English players by name. I was introduced to Italian food and their cries of dismay when I described the English dish 'Macaroni Pudding'. However, one evening I was introduced to a young Italian who told me he was working with the American OSS and was expecting to go through the lines to get into Florence before the 5th Army arrived. This lack of security for his mission was at least disturbing as he also intimated that he was also carrying private messages into Florence and had advised Rosetta of his intention and would take a message from Rosetta to her German officer if she wished. I left the flat early and consulted the FSO and he agreed that Rosetta could get out of Rome with my assistance and I was to try to get hold of any notes and then to contact the OSS. Early next morning I saw Rosetta and her sister and secured a lift to Naples. I then went back to the flat and was able to get in through a balcony window and fortunately the note Rosetta had prepared was awaiting collection and was retrieved. The note was addressed to the German officer by name only and was written in German which was quite harmless a note of good wishes and wishing him luck wherever he was. Later that morning I was able to contact the OSS Major who was responsible for the line crossing operation and gave him a report of the incident and handed over the note I had recovered from the flat. The Major, an Italian/American then 'blew his top'. He always advised his people to keep a low profile and "keep away from Goddam Italian women", he did add that with Italians some did like to impress young girls of their activities and standing with Intelligence formations. He advised me that the matter would now be handled by OSS. I told him Rosetta had left Rome, he added "tell her to stay there until the war ends" and there the matter ended.

Some months later I was travelling to Naples and took the opportunity to call in at Roccamonfina perched on top of the hill and a very rough track up to it. Rosetta was back at University but was expected home that night and I was invited to stay over. There were 10 children in the family and their pride of having Rosetta at University again was most noticeable and on arrival she had an attentive audience, and made me realise the close ties amongst Italian families. After leaving the army I met Rosetta occasionally in Naples, still a student at the University where she was concentrating on her English. Her war experiences had almost been forgotten. She did mention that she had never heard the outcome of the line crosser's activities or the German officer. When I look back I wonder what did happen and how complicated some people's lives became.

Ever Been a Tethered Goat?

Bob Steers
City Detachment, FS Vienna 1953
(formerly 291 FS)

On 22 August 1913 a son was born to Massimo Pontecorvo and his wife. They called him Bruno. It was a large Jewish family living in Marina di Pisa, Italy. Clearly, Bruno was bright for he gained his PhD in physics at Rome University and soon after went to France to continue his research work. In January 1940 he married a Norwegian girl named Marianne Nordbloom. The couple fled from the German advance into France and then to the USA and by 1943 he was working for the Anglo-Canadian atomic research team in Montreal. In 1948 he applied for and was granted British nationality and came to work at the UK Atomic Energy Authority research station at Harwell.

In September 1950 he was granted leave of absence and departed with his wife and two children for Italy, never to be seen again in the West. He sent a cheekily worded postcard to a colleague at Harwell . . . 'Goodbye everybody, Bruno'. After a short hectic period of searching the Western Powers were informed that Pontecorvo had defected to the then USSR. This was a blow as the Cold War was at its height and nuclear superiority was of paramount importance. Pontecorvo was in a position to inform Soviet scientists as to the state of our research and, to a lesser degree, that of the Americans. In fact, the Americans at that time were most concerned as to his defection. In addition to this, the Russians had yet to explode the hydrogen bomb. To this day it is unlikely we will know who or what converted him to the ideals of Communism, although we now know that many members of his family were communists. It just goes to show the importance of 'background checking'. He was a trusted government employee engaged on work of a highly delicate nature and, in terms of Intelligence,was of equal standing to that other physicist double-dealer, Klaus Fuchs.

In March or April 1953 it came to our attention that Bruno Pontecorvo was in Budapest attending a scientific congress.This was the very first time, and very unusual at that, that a defector had been allowed outside the USSR. A plan was rapidly conceived. Pontecorvo would be invited to return to us, all would be forgiven, we would ensure his safe arrival in the United Kingdom (bearing in mind that Vienna then was, politically, 90 miles behind the Iron Curtain) and in return, he would inform our authoritiies of the progress of Soviet research.

My photograph and a description of the clothes I would be wearing on a specified day in March or April 1953 were sent to Pontecorvo in Budapest together with a letter stating that should he be at the junction of Landstrasse Hauptstrasse and Ungargasse in the III (British) District of Vienna on the day concerned at 1100 hours, where he would be met by a representative of the British Government, yours truly, who would accompany him to London.

Incidentally, he was also informed that I would be carrying a copy of the Wiener Kurier under my RIGHT arm. This was an added security measure as most people are right-handed and carry newspapers under their LEFT arms and we did not want an entirely innocent passer-by to become involved in this operation.

We discussed the various reactions which could happen. They were:

(a) Presumably Pontecorvo was travelling on Soviet papers and should he be a willing participant he would have little difficulty in crossing into the Soviet Zone of Austria from Hungary as both areas were under Russian control, although there was an element of risk of which he must be aware.

(b) He could inform the AVO (Hungarian Intelligence Service) of our letter in which case they would (i) inform tho NKVD (which became the KGB in the following year, 1954) and both organisations do nothing or (ii) organise a 'counter-snatch' in order to have a pawn in their hands and to warn British Intelligence off and teaching us a lesson at the same time.

(c) A real problem was the matter of his family still in the USSR. He could be given assurances that, eventually, they would be brought to the West but, therein, was the fly in the ointment.

We could have got him out through our own methods, as easy as pie. Part of the duties of City Detachment was to interrogate IFCs (Illegal Frontier Crossers). These were those people who managed to penetrate the frontier restrictions on the Austro-Czech and Austro-Hungarian borders, such as ploughed fields, mines, barbed-wire, dogs, watch towers and itchy-fingered patrols. When these courageous souls reached the British Sector of Vienna they were arrested by the Austrian Gendarmerie and handed over to us. We would interrogate them in order to ascertain whether they were penetration agents of the Czech STB or Hungarian AVO. The interrogation would start as from the day they were born to the time they were sitting before us and particular attention was paid to the precise way and place they crossed the frontier, one, to make sure they were telling the truth and, two, for information for our own people travelling in the opposite direction. They would then be sapped of all military and industrial information they consciously or unconsciously possessed. It was surprising, with acquired skill, even an unwordly housewife had heard and seen things which were of interest to us. These interrogations would last from four days to three weeks, particularly if the 'customer' had been an engineer in an important industrial complex. We could also ask for an interrogation brief from the technical advisors in Int. Org. who knew what details were required in London.

Our cumulative corporate knowledge was considerable. We had collected, collated and passed on detailed information on everything from the uranium mines at Jachymov on the Czech-East German border (these were silver mines which had been worked for at least 150 years and uranium was only discovered after World War II, which was fortunate for the Allies) to the Czech Armd Div HQ'd in Olomouc, to the weapons factory of Zbrojovka in Brno (reputedly the birthplace of the Bren gun), to the engineering works on the island of Csepel in the Hungarian Danube (Duna), to the Soviet Armd Div in Szeged (T34s and JSU2s), to the humble Hungarian artillery unit in Kiskunfelegyhaza . . . . translated it means 'the green valley with a white house' – very Welsh. How the world changes. Less than twenty years later the Czechs and Hungarians were only too delighted to give one industrial information. On several occasions, when I was publisher of industrial journals, I have been given the red carpet treatment around industrial complexes. However, back to the IFCs. Once they had been processed they would be smuggled past the Soviet patrols on the Semmering Pass, whence they came within the domain of Field Security Styria (Steiermark). It sounds dramatic but it was quite easy and we never lost a single customer, so that's Int. Corps service for you!

As so often in life, one stubs one's toe against a situation totally by accident. One of the IFCs was a woman in her late fifties. I had taken her through her Lebenslauf (life story) but the description of her escape across the Austro-Hungarian frontier just did not add up. Those of us involved in these interrogations knew the topographical nature of most of the escape routes quite intimately and asking this woman to describe her flight with the aid of a large scale map soon made it apparent she was lying. Of one thing I was certain, she was no penetration agent but why was she lying about the escape route? Had she come the way she insisted her clothes would have been bedraggled and her shoes would have been affected by long immersion in water yet she was presentably groomed and the soles of her shoes almost new. For the next two or three hours I gently but persistently questioned her, pointing out that she clearly had information which would be of value to the West and as she had escaped in our direction, her duty was to cooperate. Most interrogation is about persuasion and convincing. Eventually, she related the true facts, undramatic but extremely important to us. Several times a week (possibly every day) a two carriage diesel or electric train would travel from Budapest to Wiener Neustadt carrying only Soviet staff officers from Hungary to their HQ in Austria and vice versa. This being so, there was no document check or halt at the frontier station of Sopron. One of the drivers of this train was the nephew of the woman sat before me. By previous arrangement, all she had to do was to go to the station in Budapest, walk on to the platform, where there were absolutely no security checks, and step into the driver's cab and stand in the hanging curtained space for his coat.

On arriving at Wiener Neustadt,where again there was no security check, all she had to do was to take the next train to Vienna, some twenty five miles away. As simple as that . . . like all good ideas. From an Intelligence point of view it was like finding the Holy Grail. I reported this information to my superiors. We had quite a traffic to and from Hungary. Thus, when our people had to return to Vienna it was a simple matter for them to make contact with the lady's nephew and 'persuade' him to accommodate them, as individuals, in the space provided for coats in the driver's compartment. As such, our 'people' were being unwittingly escorted across a hostile frontier by Soviet staff officers. You will agree, a most cooperative and gratifying act. Unfortunately, it did not work the other way round as security on the eastwards trip from Wiener Neustadt was tight and competent. This heyday only lasted some three or four months until our nephew was put on other duties and it was clearly too risky to 'convert' other drivers. However, it was during this period we could have got Pontecorvo out to us. I suggested this medium but was overruled as it was considered that we did not want to blow the de luxe travel arrangements for our people and, of course, Pontecorvo was an unknown quantity. Also, at that time, we did not know that this route had a limited life.

To return to the junction of Landstrasse Hauptstrasse and Ungargasse in the Third District of Vienna. At the final briefing it was arranged for some six or eight chaps from City Detachment to be armed and stand innocently around the junction and two Jeeps with Military Police, again armed but in uniform, to be secreted in two alleyways off Ungargasse. Should the AVO but, more likely, the NKVD attempt a counter-snatch, my colleagues would come to my assistance and the MPs would make the arrest and escort the 'guests' to a safehouse. In the meantime, we FS types would melt away and RV at the detachment flats in Sebastianplatz. As this was before the advent of micro R/T the MPs would be warned by hand signals; rather clumsy but it was the only method available at the time. Should Pontecorvo arrive we had made arrangements that he would be in London that evening by courtesy of the RAF. During the final briefing the man in charge said to me, "I don't want you to carry your gun in case things get messy". The illogic of this statement was totally beyond my comprehension. My precise answer was, "As I am to be the tethered goat I shall be carrying my Beretta and if you've any hard feelings we can always change places". This seemed to settle the matter in my favour. That night I mused that if a counter-snatch were made and if it were successful whether I would end up in Wiener Neustadt with the Ivans or with the notorious Hungarian AVO at 64 Andrassy Ut in Budapest. I say notorious because they had a bad name from their interrogation techniques, so bad that after the October 1956 Hungarian Revolution the HQ was moved across the River Danube to a new address in PEST and even the name of the street was altered. By 1969 the self-same building was used as an office for their chemical engineering industry and I had occasion to visit it for a business discussion. I had a strange feeling of deja vu, particularly as I was familar to the layout of the building.

The Pontecorvo RV was for 1100 hours the next day and we were all in position a good hour before. Near the junction and at the lower end of Ungargasse was the Cafe Nordpol. I went in to have a 'Mokka' and to make sure there were no chaps with high cheek bones and snow on their boots. After exiting, I took up my allotted position, looked around and saw the others from City Detachment scattered comfortingly around.

We waited . . . and waited . . . and waited.

Bruno Pontecorvo died in hospital at Dubna, Russia on 25 September 1993, aged 80 having been given the Order of Lenin on his 70th birthday and made a member of the Order of the October Revolution for his work to the physical sciences.

In the words of R. Burns, Esquire: "The best laid plans of mice and men gang oft agley".

309 FSS Berlin and all that . . . 1953

Peter Stafford-Hill

The only memory of Bad Salzuflen that I have was learning my first bit of German, a very useful string of oaths from some lumpen ex Wehrmacht type. The overnight trip up to Berlin made newcomers realise just how deep it was inside 'enemy' territory. Later we were told that there were about 20,000 Russian tanks surrounding us, but: 'Don't worry they wouldn't dare tackle the RTR's eight Comets.' It was this feeling of dangerous isolation which made Berlin the exciting place it was in those days. Even its insignia (the flaming a....e) reminded us that we were on our own. There was always the imminent threat of another Russian invasion versus the Berliner's determination to survive and enjoy life. It was if you like, a disaster waiting to happen at any moment. This could be seen in the two halves of the city . . . the ruined, depressing yet menacing Russian Eastern Sector compared with the lively, bustling and prosperous Western Sector with its Ku'damm, Kempinski and night clubs.

Tell me, 3 Company, does Dicke Heinrich's little cabin still exist on the Kurfurstendamm? You used to be able to get a smashing goulash there at four or five in the morning after a heavy night out: no goulash has ever tasted quite the same since.

There was no 'Wall' in those days, just a pretty lethal border over which, nonetheless, determined and ingenious refugees (plus other 'guests') continued to flood into our reception centre at Marienfelde. I found 309 Field Security Section in a private house on Allemanenallee at the time (late 1953). A month or two later we moved to an office block just outside the perimeter of the Olympic Stadium Headquarters. The section consisted of three detachments. They were: A Detachment – investigations and interrogations; B Detachment – security of civilian employees; C Detachment security of installations/information. Starting on B Detachment I found myself 'interrogating' German civilians by endlessly repeating questions from the Fragebogen: 'Was ist Ihre Name? ' . . . 'Wo wohnen Sie (liebchen)? ' . . . 'Was machst du heute Abend? . . . und so weiter.

It would have been boring had it not involved so many interesting people. There was, for instance, the ex Fallschirmjäger who had been involved in Mussolini's rescue from an island prison. Later in the war he had been taken prisoner and found himself working on a Cambridgeshire farm. To cut a long story short he had married the farmer's daughter and 'captured' the farm. He was now back in Berlin (1954) to see some sick relatives and couldn't wait to get back to his home in England and leave me in Berlin!

C Detachment came next. It was a distinctly dodgy business breaking into the HQ and various barracks. If you managed to avoid the GSO with their snarling alsatians you ran the even greater danger of being caught by any member of the Guards regiments. Apocryphal tales were told of a very painful body search that they conducted on one hapless colleague, mostly with a bayonet . . . it still brings tears to my eyes! We were known, you see, for the nasty habit of taking any classified documents left lying around and leaving a note telling the culprit that they could be collected from their CO along with a 'fizzer'.

Last of all came A Detachment, run by the late Staff Sergeant Maxie Pemberton. This was the fun stuff, interrogating returning deserters mostly (plus a few other tricks). Maxie soon showed me how to reduce them to gibbering wrecks. The Russians used to send them to live and work in Bautzen down on the Czechoslovak border. It always surprised me just how much they knew, without even realising it, when they made their inevitable return. Through them we managed to get to know 'Boris', our opposite number, quite well.

My regular job, however, was to vet all the German girls going to our Naafi dances; now that was a fascinating job. There were other duties of course, in which the whole section was involved. Spandau Prison, at the time, contained the German equivalent of the Magnificent (?) Seven: Hess, Doenitz, Speer, Schirach, Funk, von Neurath and ? (answers please on a postcard). Funk and Doenitz both went into the BMH for an operation. So there we were, fingering our loaded .38s, looking at Doenitz and thinking about all those equally helpless merchant seamen that his U boats had slaughtered in the war. Funny old world. We even got a small mention in Speer's diaries later on.

Then of course there were our regular mess guests, the delightful Gisela and her friends Inge and Jutta. Wonder what happened to them all? Socially we all shared the same mess since we wore civvies most of the time. The rate of exchange was over 12DM to the pound (or was it BAFVS?), and a packet of 20 Benson and Hedges cost today's equivalent of 6p. This meant of course that we could have a whale of a time for a few quid.

There was Dietrich Fischer Dieskau in Tannhäuser one night, Dizzy Gillespie and Ella Fitzgerald at the infamous Sportspalast another, dining in the Funkturm, riding in the Grunewald, sunbathing on the beach at Wannsee etc. Newcomers were usually initiated at the Golden Hufeisen where they were never quite the same after discovering that their gorgeous dancing partners were actually transvestites. One member of the Section (who shall remain nameless) actually performed on the stage at the St Pauli nightclub, but I shall not tell you what he did. Suffice to say, he won the competition, delighted the audience and kept up the Corps' reputation for excellence!

The average Berliner was a stout-hearted, fun-loving sort of chap, unlike his more humourless countrymen in the Western Zone. We got on rather well with them, particularly at Christmas time when those of us not on leave would be invited into their homes for a sentimental get-together. My host was an ex ack-ack gunner who had spent his war in Hamburg potting at our chaps bombing him. Nonetheless we won hands down in the friendship stakes after what the Russians had done to them when taking Berlin.

Looking back I enjoyed every minute. It is the esprit de corps that I miss most of all. Civilian life is rather like being on your own in a treacherous jungle; all its sunny clearings are full of booby traps and ambushes.

317 FSS in the Airborne Landing across the Rhine

Anon

When 6th Airborne Division returned on the 23rd February, 1945, from the Ardennes battle, into which they had been launched at a moment's notice (and as a result of which they missed their Christmas dinners and all-the rest of the seasonal festivities), many a tough paratrooper was heard to remark feelingly that surely to so-and-so the powers-that-be would let the Division alone a little while in peace. But it was not to be: and hardly had we finished our disembarkation leave when the planners got to work on 'Operation Varsity', as a result of which, one fine day in March, we were to be dropped over the Rhine right in the middle of the "promised land".

The operational plan was that Commandos were to take Wesel in the early hours of the morning, to be followed up by the infantry, and that after all that was well in motion, airborne troops were to drop a few miles beyond the Rhine: US Paratroops in the neighbourhood of Wesel itself, and the 6th Airborne Division to the North of them, in the area of the little town of Haminkeln.

The function of Field Security in the operation was to some extend circumscribed, inasmuch as no definite CI targets could be given to us, as there were none in the area concerned. It was almost entirely agricultural in character, and there was a regrettable absence, according to the information available to us, of Gauleiters, Kreisleiters, Ortsgruppenleiters, Gestapo, SD and Abwehr types, all of whom normally constitute the legitimate prey of FS. In the circumstances therefore we were given the task of dealing with the civilian population in the Dropping Zone, and our task was to see first, that civilians did not impede the course of the operation, and secondly, as far as was practicable, to screen them, in case any notorieties should have attempted to go to ground amongst them. To help with the first task some thousands of leaflets were printed in German, containing instructions to the civilians as to what they were to do. They were instructed to make out a list of all people residing in the house, to declare any weapons or 'kriegsverwendbares material', and warned that any assistance to members of the Wehrmacht would be punished instantly with death. As events proved, these pamphlets were immensely useful, and saved FS much work of a routine nature.

A very important problem was faced when a decision had to be made as to the number of FS who were to take part in the actual drop. On the one hand, there was the melancholy lesson of Arnhem to be faced: when an entire Airborne FSS was risked, with disastrous results. But on the other hand, it was obvious that in view of the tasks devolving upon FS in the forthcoming operation, to send too few would be as useless as sending none at all. After much consideration, it was finally decided that half the Section should take part in the Airborne landing – the FSO and eight NCO's and that the rest of the Section under the CSM should follow with the sea-tail (as those elements of the Division which travelled to Germany by land and sea were termed).

And so, some five days before D Day, the two halves of the Section took leave of one another. Early in the morning it was: and so the FSO, like all FSOs, was still in bed, and felt rather like a corpse already as the CSM and the other non-operational members of the Section filed past the bed to wish him good-bye and good luck. The CSM then departed with his half of the Section for Tilbury, and shortly afterwards the FSO and his party left for the marshalling area.

This was in Essex, near to the aerodrome from which we were to fly, and for some five days we were, of course, completely cut off from the outside world. This did not mean, however, that we FS were idle. We had to see to the security of the briefed troops (which included ourselves, of course!), and this was quite a full-time job in itself. There were one or two minor scares, as there always are, but the security of the operation was well maintained, as will be seen subsequently from this account. One unfortunate official belonging to the telephone service of the Post Office was about our only victim. The night before we were due to fly off, owing to an oversight, he was allowed into a room to mend a phone: and the room was one into which he should never have entered. There was only one thing to do; and we did it. We sealed him that night in the camp with the rest of us. The poor man was nearly in tears. What worried him was what his wife would be thinking when he stayed out all night. When we advised him to tell her the plain facts of the case, he replied that his wife would take a lot of convincing that he hadn't been out on the loose, and was spinning a tall story to cover it up. At the finish we turned him over to the sergeant's mess, and that solved the problem, for when we next saw him about midnight, he was as tight as a lord, and was heard to declare that he was contemplating taking a parachute course !

The night preceding D Day was glorious. The stars shone out of a clear, serene sky: and as one lay on one's bed, and regarded them out of the window, the thought inevitably came – "I wonder if I shall be seeing them this time to-morrow night?". There wasn't much time for such morbid fancies, however, for we had to be up at two in the morning, in readiness for the take-off at eight. Of the nine FS who took part in the drop, the FSO and Cpl. Hughes went in one glider with Div. HQ, Cpl. Stagnetto went in another glider also with Div. HQ, and of the remaining six, two went with each brigade, and of these four were to jump (Sgt. Miller and Cleator with 5 Para Bde., Sgt. Wells and Cpl. Edwards with 3 Para Bde). The remaining two (Cpl. Caudell and Cpl. Howse) went by glider with 6 Air Landing Bde.

D Day dawned climatically perfect; that cheered everybody up considerably, and after we had all had a cup of NAAFI char plus the traditional wad, we climbed into our gliders or Dakotas as the case might be, quite chirpily. The journey up to the Rhine was quite uneventful – the only excitement up to then had been the juction, above Brussels, with the American Airborne forces, who had flown from airfields near Paris, and who joined us above the Belgian capital by previous arrangement, and who then flew side by side with us on the final stage, the most exciting!

The Rhine looked insignificant from the height at which we were flying; but we soon knew we had crossed it when the flak started coming up at us. At last came the great moment: the glider pilot told us we were about to cast off, and then we started the last stage of our journey, and began to glide to earth. We had a fairly comfortable glide, though it was rather eerie swishing down through the artificial fog and waiting all the time for the light flak and Spandaus to open up. We were lucky, however: US paratroopers had landed a few minutes earlier in our zone, and so by the time we landed quite a lot of the opposition had been mopped up.

We made a perfect landing, and after a few moments during which we sorted ourselves out and took our bearings, we set off for our objective – a group of fortified buildings.

Here again we were lucky, for once more the Americans had got there before us and dealt with most of the opposition.

Having installed ourselves, we thought that we might at last be able to get on with some FS work: but it was not to be. Germans surrounded our position, and kept us under continuous fire, mostly small-arms, but later on with mortars, too, to liven things up.

In these circumstances Field Security work was out of the question, so the FSO became perforce a platoon commander while Cpl. Hughes performed invaluable work in attending to the wounded under heavy fire. After about seven hours of this siege we received orders to withdraw to main Div. HQ which was about a quarter of a mile away, and this, we were able to do without casualties, though the withdrawal had to be effected over open country under continuous enemy fire. At Div. HQ we were glad to find Cpl. Stagnetto who, like us, had had no chance to do any security work, but had had all his time occupied in going for the Boche.

Already, however, some hundreds of frightened civilians had found their way by one means or another, to the comparative shelter afforded by Div. HQ, and so we got to work on them, scrutinising identity cards, and generally going through the whole FS bag of tricks. The biggest headache was the problem of feeding and accommodating these people, for amongst them were babies, old and infirm, and cripples. However, somehow it was done, and although we says it as shouldn't, it was a good job FS were there to see to it, for otherwise a very grave problem would have arisen which might have had to be dealt with drastically in the circumstances then prevailing.

Having sorted out all our civilians, and left them under guard, FS proceeded for the rest of the night to man slit trenches, and the afternoon of the following day had the pleasure of contacting four more colleagues who had come on the operation. That left only Sgts. Miller and Cleator unaccounted for; and for one more day we were left in suspense fearing the worst, but hoping for the best, till our fears were set at rest by the appearance in person of the two missing NCOs. They explained that their Dakota had been hit by flak over the Dropping Zone, and that the pilot had decided that it was impossible to come down to the low altitude necessary for the jump; so they had flown back to England whence, after a short halt, they had returned by air to Eindhoven, and then by road to us.

Thus once again all the Section was together, and (though we did not know it at the time), with two months of interesting and exciting FS work in front of us. But that, of course, as they say, is another story!

PS. Airborne Div. later captured the German general commanding the ground troops who opposed us when we dropped; and in the course of conversation he admitted that, though he had a good idea of where we were likely to drop, the actual time of the operation caught him completely napping. So there's something in this 'ere security racket after all!

317 (Airborne) FSS 1947

John Short

The Section was commanded by Capt. John Linklater MBE, one of the most fascinating men I have met in my life. His mother was Czech. and his father had been the Senior Master at the Slade School – not THE Slade – in West London. He was an old Pauline of, I am assured by a schoolmate who hated him, exceptional brilliance. John Linklater spoke his mother tongue, Czech and ". . . if you can speak Czech you can speak Polish" (his own words), French, Italian, German (having twice been in German hands) " . . . some Russian". He was very much an individual and it would not be an exaggeration to say that he was hated by most of the officers on the staff of 6th Airborne Division. If I may relate an instance of this when the General Service Medal with the Palestine Clasp was awarded for active service in Palestine only two officers on the divisional staff were able to 'put up' their third row of medal ribbons. One was the GOC Maj. Gen. (later General Sir) Hugh Stockwell and the other was John Linklater who held two Czech decorations (with the King's permission) one of which was the Czech Medal for bravery, described by him to me as 'Virtuti Militari'. I think he was respected by Gen. Stockwell in much the same way that Wavell respected Wingate.

He was totally unorthodox and revelled in it. He was capable of beating up Jewish leaflet stickers, when caught ". . . I do not believe in torture but a good beating-up . . ." (he had himself experienced torture at the hands of the Germans and, I think the Italians and Spanish as well). He was full of what were most – but not always – practical jokes such as getting all of us out of the Jeep which driverless, we would then accompany – it being in 4-wheel drive and 4th gear – through the alleys of the 'Soukh' in Haifa. But he also took the same Jeep up the Prophets Steps which formed a link between the Arab area and the Jewish area of Hadar Hacarmel which caused the Haganah to put a much stronger post on their end of the steps. When the Pan Zephyr and the Pan York were intercepted and taken to Cyprus, half of 317 were sent over to reinforce the local section, and because we had become specialists in interrogating IJIs (Illegal Jewish Immigrants). There, amongst the crew one of the vessels was a Britisher who made the mistake of pretending he was Czech. The result was questions in Parliament coming down the line which I, who had by then taken over, was obliged to field.

There is no doubt in my mind that he was more than just an FSO. He established a good working relationship with the Haganah in Haifa and I am sure that he was operating under some command different from that of 6th Airborne. He left the running of the Section more and more to me and became increasingly a night owl. So much so that one of my duties was to make sure that he got up in the morning which increasingly I left until as late as possible and on one occasion decided that he really needed the rest and let him sleep which made him extremely angry. But I subsequently learned from one of my sergeants who accompanied him on these nightly excursions that ". . . you had no idea what was going on . . ." and that on several occasions they went across Palestine into Trans-Jordan as it was then. Towards the end, extremely tired, I approached his quarters to waken him and found the GOC and the GSO 1 there before me. I was told to wait outside and when they emerged the General said to me "You are now in command, John is leaving on the troopship tomorrow, see that he gets on it". I was not yet 21 and although I was learning I was totally unqualified to follow such a man.

Thereafter began a different period and as we were withdrawing from Palestine, Gen Stockwell appointed me as his 'Jewish Liaison Officer', and Tom Keen (the OC of the Static FSS) as his 'Arab Liaison Officer'. Our job was to round up the respective leaders when the General wished to have words with them. Very many years later I met again the man who was (although I did not know it – I thought he was one prominent Jew) the then leader of the Haganha, Jacov Salomom, I was paid what I consider was one of the greatest possible compliments when his wife said to me, "I remember you well, when you came for Jacov we knew he would be safe".

There are many stories connected with that period in Palestine such as the learning of a few words in Hebrew from Harry Bewlin who was my principal contacts with the Jewish Agency and being in civilian clothes (as we sometimes were) using them at the Jewish barricades, which protected the Jewish areas of Haifa until the young men manning them thought that I looked a bit too young to be the son of the Rabbi of Liverpool. I do not wish to include all the many incidents which remain almost branded on my mind for I truly 'came of age' at that time. I fear that some of the best sources are now gone, like John Linklater, but there was another FSO Ernie Leeves whom I met on the first trooper coming home and who was another of the great 'characters' – First World War ex-gunner, ex-Black and Tan, ex-Palestine Gendarmerie (Palestine Police), ex-behind-the-lines road-watcher in the desert, fluent in Arabic and Hebrew, bottle-of-gin-a-day man who was still lean as a whippet and whom an entirely independent observer who had known him in the campaign in Greece after the war proper, told me could run up a mountain quicker than men half his age.

Much of the work of the Section was that of a normal FS section – security and property, security of information, etc. In as far as it related to 6th Airborne Division as there was also a 'static' section based in Haifa. But because 6th Airborne was the major unit in northern Palestine our GOC was in overall command. 317 FSS was housed in the underground offices on Mount Carmel previously, I believe, the RAF HQ, but by 1947 we were the sole occupants. There was a detachment with 3rd Para Bde. at Nazareth and for a period of time one Sergeant on his own at Safad. The Section was fortunate in having its own 'mess' which was the duty room manned 24-hours a day but the sergeants and corporals were accommodated in the officers' lines and were members of No. 3 HQ Officers' Mess.

However, and I think partly because of John Linklater, we were heavily involved in counter-Jewish Terrorism. This included trying to recruit informers, anti-postering sweeps and latterly liaison with the by then virtually recognised Haganah. We also had the often unpleasant task of interrogating the human cargoes of illegal Jewish immigrants whose ships had been taken over by the Navy when they attempted to land in Palestine. On only one occasion did an IJI ship get through, as British and Naval intelligence was normally very good. The normal practice was for the ships to be intercepted by the Navy, boarded and taken over by the Royal Marines and then brought into Haifa where all but the very old, the sick and the heavily pregnant were transhipped to the camps on Cyprus. We normally knew if the transhipment would be an easy or a difficult one. If a baby had died during the voyage it would be displayed through a porthole for the benefit of the world's press and there would be no resistance, but on other occasions there could be quite a struggle.

Once the ship had been secured it was the task of the Section to interrogate the IJIs as they passed through the disinfecting 'pens' either on the stern of a cruiser (in the early days) or on the docks at Haifa. Other members of the section would search the vessel for documents and this was a most unpleasant task, as the stench (which was deliberately created about one day out – we had photographs which had been taken on board which showed a different state), was terrible and also the IJIs knowing that the ship would be searched, often concealed 'ordure' amongst papers left in drawers: at least they were not explosive booby traps!

It was John Linklater's boast that within 24 hours of a vessel's arrival a report giving details of where the 'cargo' had assembled, where the boat had come from and how long they had been at sea plus any other information would be in GHQ in Jerusalem. Me, being more or less a spare, had the job of being the scribe to his dictation (once or twice with the aid of Benzadrine).

We also had to turn out for any major incidents such as the blowing up of the Police Station at Haifa and the massacre of the Jewish clerical staff at the Iraqi Petroleum Company outside Haifa (we rescued three survivors).

Two FS sergeants of 317 FSS were hung by the IZL, possibly in May or June 1947. This was an abbreviation for Irgun Zvani Leumi (I am not sure of the translation from the Hebrew but Leumi is 'people', if my memory serves me correctly). They were a radical right-wing organisation and favoured dramatic terrorist activity. While the majority of the Jewish population at that time were 'moderate', the

IZL were not, although subsequently a number of their then leading members became very powerful in Israel; Menachem Beigin was, when I was in Palestine, the most wanted man on our lists and he went on to be Prime Minister.

I myself was with the Section in Port Said at the time when this happened but shortly afterwards I joined 317 Airborne FSS in Haifa. Two of the sergeants on that Section had handed over in Natanya to the two murdered sergeants some months before when the Airborne Division had moved to northern Palestine and were replaced by the Guards in central Palestine. It was the opinion of these two sergeants, who had been operating in fairly similar conditions, that the two murdered men had been extremely foolish and had allowed themselves to be so far influenced as to become pro-Jewish. Certainly the local Haganah in Natanya had done their utmost to find the men but had been frustrated by the imposition of a curfew by the British forces, although the Brigadier i/c of the area, had appealed unsuccessfully to Jerusalem to allow the Haganah more time to obtain the men's release; there is no doubt that there was considerable revulsion in moderate Jewish circles about this.

It was not uncommon for the Haganah to act against the IZL or the Stern Gang, although they would often knowingly allow them to carry out 'terrorist' operations when it suited Haganah's own purposes to do so. On one occasion, to my knowledge, when IZL had planned an operation in Haifa at a time when Haganah felt it was in their interests to foster relations with the Airborne Division, the IZL team coming up from the south were met on the coast road, beaten up and sent home 'to think again'.

It is not always so, but 317 FSS undoubtedly made friends – as was part of our task – with the Jewish authorities, both legal and otherwise, in northern Palestine. We were disliked for our involvement as the interrogators of the illegal Jewish immigrants whenever their ships were intercepted and brought into Haifa, but with the local community our relations were good. On one occasion the IZL threw a bomb into the crowd of Arabs waiting outside the IPC refinery just outside Haifa, who were there waiting for casual employment, myself and three sergeants were sent down to find out what had happened. What had happened was that the Arabs outside the refinery had called out to their compatriots inside the refinery to avenge them, which the mainly Arab labour force had proceeded to do towards the mainly Jewish clerical/technical employees. We did find two of these alive and in hiding and brought them out through a mob of hostile Arabs (justifiably so) which was a very 'hairy' experience.

We were on 'friendly' terms with Jewish 'civilains' whom he had identified as part of the Jewish Army about to be. It was in fact in being and we used regularly to visit the man who was the 'battalion commander' for the area of Haifa docks. Our last visit to him was when he was in hospital after the Arabs had managed to deliver a vehicle bomb (driven by German ex-PoWs) into the harbour area. It was not a sophisicated bomb and had a trailing, lighted fuse showing, which he had tried to cut by shooting at it, when the bomb went off. He had served with Wingate's night squads pre-war and when he asked him what would happen to him now he replied that he was finished, as the Jewish Army did not take kindly to failure.

326 FSS in Germany 1945–47

Tom Salmon

In August 1945, I left Bad Salzuflen, where I had been attached to HQ Intelligence Corps (Field) in order to attend the School of Military Intelligence at Paderborn. This was part of my transition from the Infantry, in which I had been serving, to the Intelligence Corps. Previously I had been tested as a German linguist by Major Sedgwick Rough.

At the School of Military Intelligence we learned all the arrestable categories of the Nazi Party machine (SS, Gestapo, Sicherheitsdienst, Abwehr, NSDAP etc.). There was also a background of German politics and constitution to be absorbed.

At the end of the course I made a brief return to Bad Salzuflen, where I received promotion to the rank of sergeant. I was almost immediately posted to 326 Field Security Section in Kiel, and arrived at the Section in September 1945, to find myself with the questionable honour of being the youngest member of the section. The OC was Captain Whitley, soon to be replaced by Lieut. Morrison.

326 was a Port Section, and, as such, we had the permanent use of a sleek motor launch which we used daily to ply the Kiel Canal and harbour, boarding and searching all the ships which came through the Canal, of diverse nationalities. The searches were for the

arrestable categories of Nazis mentioned above. One of the first sights I saw in the harbour was the famous German warship Admiral Scheer, upside down and an obvious victim of Allied bombing. There were many and varied wrecks in the harbour, not all as famous as the Admiral Scheer.

Much information came to me at this time from Heinrich Erichsen, an anti-Nazi who was a pilot on the Kiel Canal. Erichsen had been a Communist and, as such, had been physically badly knocked about by Nazi thugs in the Hitler era. He was particularly knowledgeable on the subject of the former Grenzpolizei (frontier police) on the docks at Kiel Holtenau. Erichsen was a fine man with a broad international outlook brought about by his work with ships of all nationalities and he remained my friend until his death many years later.

During the weeks that followed, apart from the work on the Canal, I was also engaged in what was known as 'Shore Coverage', in which we interrogated the crews of all the barges and the huge floating cranes around Kiel harbour. My colleague on these missions had hardly any knowledge of German, hence I did the interrogation, but he was a fearsome figure who had been through the Desert campaign. He must indeed have struck terror into guilty consciences without a need for words!

At Christmas 1945 I went home on ten days leave. I travelled, in mid winter as it was, in a train with all the windows bombed out, and wooden seats, through a freezing cold night and day to Calais. The train, for a large part of the journey, only travelled at a few miles an hour, as the tracks had been so badly bombed. Owing to the shortage of coal, the engine burned wood and frequently stopped, having run out of wood! We tried to put our groundsheets up at the train windows to keep out the worst of the cold.

At last, we arrived at Calais, only to be told that there were storms in the Channel and we would have to wait 24 hours another overnight stop. I did eventually get home in time for Christmas!

Back in Kiel, the port and harbour work went on, through a freezing cold winter. The interrogations were extended to what remained of the great shipyards on the East Bank, though the whole of the East Bank had been virtually flattened by allied bombing, and only skeleton labour forces remained. For transport we had Jeeps, 15cwt trucks, requisitioned cars and utility vans. The Jeeps were open, hence extremely cold in that severe winter. Kiel was a ghost city, with many people living in and under the rubble of the bombing. There was little food, fuel or electricity, but, amazingly, there were no epidemics.

Early in 1946, I was summoned to Hamburg for a month, to interrogate, with other German speaking NCOs, the crew of the German battleship Graf Spee. The Graf Spee had been scuttled off Montevideo after the Battle of the River Plate in December 1939, one of the most famous naval battles of the Second World War. Her Captain, Captain Langsdorf, had scuttled her and then committed suicide. The Crew had been interned in Uruguay for the duration of the War and had just now been returned to Hamburg.

Returning to Kiel, the port and harbour work went on, but I then had a spell relieving on a Danish frontier Section at Flensburg, and also a spell in Lubeck.

After a while, as the Port and Harbour operation folded up, I was posted hundreds of miles away to a frontier section at Kaldenkirchen, near Krefeld, on the Dutch/German frontier. There the work consisted mainly of manning the frontier post and searching vehicles such as vegetable lorries for concealed fugitives. I was allocated a BSA 500 motorcycle, as well as the use of a 15cwt truck.

But I had not seen the last of duties in Kiel!

In the summer of 1946 I returned to Kiel, this time to 100 Intelligence Team, which was housed in a requisitioned villa with its own lake. The villa had belonged to the German margarine factory owner Harmsen. The work had changed a great deal. There was no more port and harbour work. Arrestable categories of the SS, Gestapo etc. were still sought, but the emphasis was also on the screen of civilians for the new Administration which would, hopefully, be the foundation of a new and democratic Germany. In this connection, a massive new questionnaire called a Fragebogen had been introduced to try to exclude ex Nazis from influential positions. If there was doubt as to the completion of the Fragebogen, the person concerned would be called in and interrogated.

During this period, I interrogated ex-SS and concentration camp guards, often taking those concerned to the internment camp at Lübeck, about 40 miles distant, by road vehicle, for trial by military tribunal. Some of the interrogations I did were very severe and intense. I especially remember the interrogation of an ex-concentration camp guard from Stutthof camp, near Danzig, who protested that he was as innocent as a newborn babe.

There were complaints by Jewish people who complained of past persecution, and these had to be investigated. Quite often, there was a midnight raid to be carried out on a city apartment or a lonely farmhouse, looking for wanted people. Sometimes two or three of the Team went together, sometimes I went alone with a German policeman who was allocated to us (I still hear from him, at age 87, today). Sometimes we went into the Displaced Persons (DP) Camps on enquiries. The atmosphere in these camps was quite explosive, and there had been several instances of lynching of the civilian population by the Poles, who were out for revenge. On one such occasion, I walked between the rows of huts with a particularly exuberant colleague who made a habit of swaggering along with his hand on the holster of a massive revolver which he strapped at his side. One could feel the tenseness in the air and the walk seemed endless. I wished for a more quiet and level headed colleague. But we survived!

I must admit to having 'bent the rules' from time to time. On one occasion, when I could not furnish proof against a former Gestapo/SS man I found another way of meting out a measure of justice. I found out that, in the midst of the famine and shortage, he had a full coal cellar. I took a companion and a 15cwt truck one night, loaded a goodly quantity of his coal, and delivered the coal to the impoverished widow of a concentration camp victim.

I was made responsible for the 'screening' of the City of Kiel Police Force – a very sizeable force. The Chief of the Uniformed Police was named Lietz, and he, a very entertaining character, had been in Oranienburg concentration camp. He made my job easier. Checking on one Kriminal Polizei (CID) Officer, I contacted the Berlin Documents Centre for information and found that the same man had been an SS officer in the Balkans campaign. He did not wait for further investigation, but committed suicide.

In the work of preparing for a democratic Germany, my contacts included the trade union representatives of all the shipyards and factories. These men had mostly been members of the pre war SPD (Socialist Party) and had clean political records.

The winter of 1946/7 was extremely severe in Germany, The harbour at Kiel was frozen solid for a considerable time, and snow was piled feet high everywhere. Our vehicles had to be drained off each night and refilled the next morning – these were the days before antifreeze! Food, coal and other fuels, and electrical power were all scarce commodities. I well remember this, because it was at this time that I contacted a rather dangerous mouth infection, with poison spreading up towards my nose. I was operated on by candlelight, with the German dentist's wife holding the candle, and without any proper anaesthetic. I survived, but it was an occasion I would not like to have repeated!

In the spring of 1947, I was informed that a team from the War Crimes Commision was coming to our area to investigate what remained of the concentration camp AEL Nordmark at Russee.

When the mass grave was opened up, I was the only member of the Intelligence Team to be present, and I had arranged to be accompanied by a German CID (Kriminalpolizei) Officer with whom I was associated. It was a terrible experience and one I shall never forget. Over 500 inmates, who had been virtual slave labourers, had died in the camp. Many had been shot or starved to death. A large number of the prisoners had been Russians and other East Europeans. The Commandant of the Camp, Johannes Post, was arrested by our Field Security Section in Westphalia, where he had taken a false identity and was in business as a haulage contractor (details of wanted men were circulated). Post, a Gestapo Official, was condemned to death by a Military Court in Hamburg, not so much for his infamy in the Camp, but for the murder of four Allied pilots who were handed over to the Gestapo after they had escaped from a PoW. camp. Post's deputy, Otto Baumann, was also condemned to death, and a number of the Camp guards received long prison sentences.

When the War Crimes team had completed its tasks, the horrors of the camp were, at my instigation published in the Schleswig Holstein newspapers. I wanted the facts known. The Gestapo and SS had been completely ruthless in their administration of the camp and would brook no interference from any outsider. Any would-be interfering outsider was likely to become himself a victim.

Many years later, in 1983, I received enquiries from a German researcher who was writing up the story. I replied to the best of my ability, and was also able to trace for him the War Crimes pathologist who had carried out the examinations on the victims in 1947, he was was by now Emeritus Professor of Forensic Medicine at Guy's Hospital, London. Accordingly, the two men arranged to meet in London for an exchange of information.

An event occurred during 1947 which made me feel that corruption had crept into the administration of the British Zone of Germany.

It happened in this way: we received a direction from Intelligence Division authorising us to replenish our vehicle stock by requisitioning suitable vehicles from known ex-Nazis. It so happened that I knew where a vehicle was concealed in a straw barn, covered in straw, on a large estate belonging to an ex-Nazi landowner.

I accordingly set out with my own vehicle, a companion and a tow rope and, on reaching the estate, which was many miles away from Kiel, duly uncovered the car in exactly the place I had expected it to be. There were loud protestations from the landowner, but we made the necessary preparations to tow the car away. We had not progressed far on our return journey when a Jeep came hurtling from the opposite direction, full of Special Investigation Branch men, bristling with guns. It transpired that they had been despatched by a high official of Military Government in Kiel who must have been a friend of the ex-Nazi landowner, who had telephoned him while we were uncovering the hidden car. In spite of all their bluster, there was not a thing they could do, as I had my copy of the authorising directive. They drove off back to Kiel, and we towed the requisitioned vehicle back to our unit.

However, when I returned to Kiel, I found that there was a warrant out for my arrest, issued by the Military Government official. It was a difficult time for me, but my forceful chief, Colin Mayhew, civilian head of No. 1 Sub Area Intelligence Office, backed me to the hilt. The verbal tussle was short and sharp but the outcome was never in doubt. The arrest warrant was rescinded and the requisitioned vehicle remained firmly ours. A point had been made. But the feeling of corruption in the Zone lingered.

It was at this time that I heard, through a German contact I had cultivated at Kiel Friedrichsort, of a potential case which interested me very much indeed. I told my Commanding Officer that I intended to go ahead and investigate and that if I found the necessary evidence, and could not interest our own authorities, I would turn the prepared case over to the German civil courts for possible prosecution. My CO shrugged his shoulders and gave me carte blanche to go ahead. The facts as I remember them 47 years later, were as follows:

One evening, during the war, a small group of men had been playing cards at a house in Kiel Friedrichsort. On the wall of the room in which they were playing hung a picture of Adolf Hitler. At some point in the evening, after a few drinks, one of the men, who was Director of the local firm Poppe AG, and who was named Brauer, looked up and pointed at the picture and said distinctly "That man is a madman. He will bring us all to destruction."

Unknown to Herr Brauer, one of the men present, named Durschner, was an informant of the Nazi Secret Police (Gestapo).

Durschner made it his business soon afterwards to report Brauer's statement to the local Ortsgruppenleiter (Nazi Party leader), a man named Struckmeyer. He, who could have kept quiet if he had been so inclined, did the opposite and took the fatal step of reporting the matter to the Gestapo.

Brauer was arrested and brought before the Volksgericht (People's Court) in Berlin for a show trial. He was, of course, convicted, as was every anti-Nazi brought before the so called "People's Court". Brauer was, in due course, hanged from a meat hook.

This case caught my imagination and, during the following weeks, I tracked down the people concerned and interrogated them at our office in Kiel Duppelstrasse (which had been, coincidentally, the offices and cellars of the Gestapo!).

I cannot remember a great deal of details of my investigation but I remember that I was conscious of not having sufficient evidence against Durschner, and so I made him come in regularly for interrogation, which entailed him in a long and arduous march from Friedrichsort!

When my evidence was as complete as possible, I handed the dossier to my friend Dr Paul Thamm, the Public Prosecutor for Schleswig Holstein. The case was not heard in the courts until after my demobilisation in Oct. '47, but shortly afterwards I had the satisfaction of being sent the report of the court case by Dr Thamm, and learning that Struckmeyer had been found guilty of Crimes against Humanity and jailed for several years. The Gestapo Agent who had arrested Brauer was dead.

Such was the state of affairs in Nazi Germany, that a careless remark could mean death. An interesting footnote: Brauer's wife had been a friend of Frau Goering (wife of Reichsmarschall Hermann Goering), but that did not save Brauer from the hangman.

402 FSS
The Russian

Jack Tatham

The Iraq convict sharpened old memories. Old men forget! Distances in that country were hardly mentioned in the recent communiques. only aircraft flight times. Baghdad to Basra is almost 350 miles and in 1942 I was posted, a solitary Field Security sergeant, to Samawa, 170 miles south of Baghdad, about half distance. "Go spy the land" was my brief and "bring us all the relevant information on the lines of communication, rail, road and the River Euphrates, between Baghdad and Basra", no easy task for a poor Arabic speaker who had just completed his first Security Intelligence course.

One morning in 1943, after routine visits to usual local contacts, I waited for the Basra to Baghdad train to arrive at Hilla, having earlier received a coded signal by the railway's telegraph line. A section colleague, Sergeant W was en route to the capital escorting a suspect Russian far from his homeland and my help could be needed.

Sergeant W left the train even before it had stopped at the station platform and in such an agitated state he was almost incoherent. "The swine broke the toilet window. only about two feet by 18 inches, somewhere near Diwaniyah and jumped the moving train." His language was choice, unfit for my reader's tender ears and he rambled on about his coming court-martial and other military horrors. I could visualise him smashing into that toilet door when the Russian was suspected of prolonging his sojourn there, and the likely troubles I would have in the future with Iraq State Railways – being the only security man with an imposing pass valid for unlimited travel on that route. HQ was forever reminding me that the pass cost the Army some £50 monthly – yet I had never been asked to produce it once!

Calming Sergeanl W, I accompanied him to Baghdad on the same train and reassured him that the Russian could not get far from the railway because back at the office I would soon have 'the dogs' onto him; in the meantime he should makc himself scarce so no questions could be asked of likely ETA of the expected Russian from Basra. That late aftemoon at Section HQ was spent in surreptitiously telephoning those of my contacts in Southern Iraq likely to be of some help.

The first approach to Wg Cdr C Area Liaison Officer, Combined Intelligence Centre Iraq (CICI), at his house in Hilla drew a blank: he was out for two days visiting local tribesmen. The Arab clerk to Major M, Assistant Political Adviser. Diwaniyah, was my second call. He owed me a favour and responded immediately. I was assured that if the Russian was anywhere in that entire province he would trace him that very evening. My last contact was Capt. H, Arab head of the Iraq Railway Police covering about 200 miles of the line in South Iraq. H, who was under the mistaken impression that I had secured him his recent promotion to his new rank – and I did not disillusion him – swore that each and everyone in his force would immediately be placed on alert and search parties would explore even desert tracks towards Najaf and Kerbala, far from the railway lines.

It was a long evening's vigil at Section HQ sitting by the phones, the only interruption being agitated enquiries from Sergeant W, secreted in Baghdad town, about developments. About 2230 hours the resourceful Captain H reported that one of his detachments some ten miles from the railway west of Diwaniyah had apprehended the Russian who was safely handcuffed to a table in a village coffee house. My prompt reply was that I would fetch him back to Baghdad that night.

The Section had a speedy-left-hand drive Ford 15cwt pick-up with oversize tyres, useful in sand and off-road travel. So after covering some 80 miles with blazing headlights, about 1am the lonely coffee house came in sight. Inside by the light of a naked solitary bulb I found half a dozen unshaven Iraqi police in dishevelled uniforms with collars undone drinking arak at a table, watched suspiciously by a smartly dressed, alert Russian, about 40 years of age, solidly built, handcuffed and now leg-ironed.

Captain H drew me aside and produced a black automatic pistol like a Beretta, with hammer and sickle embossed on the butt, some papers and maps in Russian and Arabic, and a Russian booklet with diagrams showing measures to be taken against German paratroopers and irregulars. Captain H confided his view that these possessions clearly indicated the Russian, who spoke some Arabic, was a Soviet spy. We then had to celebrate the detachment's success with arak; next it was my turn to buy arak all round for the captors. Oh, the headaches of these small successes!

I unloaded the automatic and slipped it into my left-hand jacket pocket, the Russian was unshackled and bundled into the right-hand front seat beside me and the headlights picked up the track as we made back for Baghdad.

We had travelled for about one hour when the headlights picked out a white road sign, in Arabic and English, indicating 'Babylon, 5 kilometres'. Shortly after this the Ford's engine stuttered and then expired. Much abuse of the starter motor and battery failed to bring the engine to life and the Russian and I were alone on a rough desert track stuck in the middle of nowhere. As readers may well be aware, desert nights can be intensely cold and deathly quiet and the unmanacled Russian soon feigned sleep; my thoughts were also that end but I imagined my reluctant passenger was endeavouring to lift his pistol from my far pocket. I hastily left the vehicle, lifted the bonnet and by torchlight extracted the fuel filter, cleaned it with petrol from a can and after replacing it with silent prayers, the Ford engine started at my first attempt and kept ticking over.

At Hilla the YMCA had the previous week opened a large house as a forces' hostel and the surprised warden opened up at about 6am to my knocking. The Russian and I were supplied with hot coffee and a truly enormous breakfast and from there I telephoned the Section HQ. Some two hours later the Military Police were awaiting me there for the Russian.

The saga of the Russian had just begun – and continued for years. The British at first suspected he was a Soviet agent. The Russian Embassy somehow learned of his presence and demanded him as a

Army deserter from the Caucasus (their story!) who awaited execution. The Iraq Government wanted him on a charge of murder of two Iraq citizens just three days before he was first apprehended. Their report was that he had approached an Arab black marketeer in Ashar telling him that he had buried a cache of valuable stolen Army lorry tyres in the desert north of Basra and for cash a deal could be done.

The Arab had hired a taxi in Basra, the Russian sitting on the rear seat. In a lonely desert location the two Arabs in the front seat shot quickly through the head by the Russian who relieved the crooked trader of his cash and disappeared. The gun used was the one handed to me by the police after his capture.

In London in 1948. Major V.M, former GSO 3, I(b), Iraq, then at Foreign Office, informed me that when he left Baghdad the file of the Russian was over one foot thick. Readers may speculate on which of the three claimants to the body won: the Iraqis – they hanged him.

A source of satisfaction to me is that the voluminous file was not expanded by the above episode when the Russian almost got away; at least Sergeant W was always grateful for that.

Now many other Iraq memories return. The cosy frontier detachment at Tel Kotchek on the Iraqi–Syrian–Turkish border, north of Mosul and Sergeant A there. How he constantly complained to me that he had been forced by HQ to pay a small monthly retainer from the meagre 'secret' funds to the bashful village schoolmistress, who was to search intimately any suspect female traveller on the Taurus Express when he would gladly have performed that delicate duty gratis.

Then there was a small downtrodden Iraqi secret service man with his smart Sidera (hat) and shabby suit who always sat chain-smoking on the same seat at Baghdad's main railway station. Clearly not a discerning smoker, he cherished the dreaded 'Victory V' cigarettes! When supplied, he deliberately turned his back on our nefarious activities on and about the Taurus Express at crucial arrival and departure times.

There was the German Iraqi agent arriving from Turkey with his new wireless transmitter who gave the slip to some 20 of us staking out his arrival at Baghdad Station – but was caught the next day in a Rashid Street coffee shop.

Memories, too of the attractive blonde Swiss secretary, Miss P, who had worked in Kabul for the Japanese Military Attache there and whom the Section 'Lothario' extensively 'debriefed' on her arrival at a Baghdad hotel where he lived on lavish expenses for many weeks – until the CICI case officer discovered just how attractive she was and took over the case personally.

Oh, there were countless other memories too . . . but old men tend to forget!

412 FSS
Trieste, 1945:
Face to Face with the Jugs

Douglas Lyle

I was with 21 Port Security Section when it arrived in Trieste in the second week of May. The Eighth Army Section (412 FSS) was already in the city. Within minutes of taking over a bank in the city centre earmarked as our HQ a phone rang. Being nearest I took the call and heard a voice say, in haltlng English: 'We know all about you.' When I asked who was calling the reply was a warning to be careful. We took it for a some kind of prank until we found there was a Yugoslav machine-gun in a window of the first floor of a building opposite, pointed directly at our sleeping quarters.

'The Jugs' was the convenient short-hand used by the Allied forces for the Yugoslav Partisan Army which on 1 May occupied Trieste, in spite of a formal agreement not to. They proclaimed the city and the adjoining area to be part of Yugoslavia. The claim was rejected by the Western Allies. But Tito's 4th Army, which had fought its way into the city a day or two before the arrival in the city of 2nd New Zealand Division, held its ground. They complained that the New Zealanders had entered the city 'without the permission of the Yugoslav authorities'.

The 'Jugs' were a disciplined and well-organised force. They moved quickly to set up in Trieste a civil administration in the hands of Italian communists, a *fait accompli* respected by the Allies. The OZNA ('Department for the Defence of the People'), the Yugoslav state security organisation, was soon a by-word for terror among the Italian population, who contrary to the claims of Yugoslav propaganda, bitterly resented their Slav occupiers. The city seethed with talk of requisitions, arrests and abductions, and rumours of summary executions.

It was a strange and potentially dangerous situation. Almost overnight Yugoslavia had become the immediate target for Allied intelligence operations. Field Security personnel were soon aware that the OZNA, ruthless, professional and efficient, had identified their function.

Attached temporarily to 412 Section, I was warned of the absolute need to protect the security of contacts and informants. A propaganda tactic used by the Yugoslavs was the organisation of large 'spontaneous' mid-day anti-Allied street demonstrations, mounted by Italian communist supporters, complete with music, scarlet banners and choruses of the Internationale. The Section had however acquired a useful source of advance information about the place and timing of the demonstrations. The Italian communist leader, Togliatti, initially advised communists in Trieste to cooperate with the Yugoslavs but later changed his mind.

There was also a back-log of intelligence tasks unconnected with the Yugoslavs to be attended to. A search was on for a supposedly important German stay-behind agent, and for three generals of Mussolini's fascist army. All were tracked down and arrested within days – a fate they no doubt felt preferable to falling into the hands of the Yugoslavs.

The city was now garrisoned by Allied troops including New Zealanders and US forces, and by Tito's Partisan Army. The underlying political and military situation was tense, but the strains were not immediately obvious. The Yugoslav soldiers, women as well as men (there seemed to be mixed platoons), sometimes garlanded with bandoliers of ammunition, shared the streets with Allied soldiers, saluting them punctiliously, NCOs and officers alike. The men went in for impromptu circle dances, often holding up the traffic. There was a Sunday football match between British and Yugoslav teams in the Trieste stadium. It was a good-tempered game won by the Yugoslavs.

We encountered the darker side when a detachment of armed Yugoslav soldiers turned up one lunch-time outside the restaurant where security personnel habitually ate and demanded that a young civilian eating with us be handed over to them. It was claimed he had collaborated with the Germans. With exemplary coolness and determination the sergeant-major of 412 FSS confronted them, demanding how they dared to behave towards their allies in such a fashion. It worked. The officer in charge of the group hesitated, and withdrew the detachment. It was a tense episode, which could have had a serious outcome.

There were various incidents during this period which remained a mystery to me, because I left Trieste soon after the eventual Yugoslav withdrawal. A security sergeant was unaccounted for for twenty-four hours, after being an involuntary overnight guest of the Yugoslavs. He returned with a hang-over. I believe he was not the only member of the Allied forces to be entertained in this way. There was talk of informal contacts between FS personnel and members of the OZNA.

Powerful political and military pressure was now being applied to the Yugoslavs by the western Allies. On 19 May Field-Marshal Alexander ended a message to the forces under his command with the words: "We are waiting to hear whether Marshal Tito is prepared to cooperate in accepting a peaceful settlement . . . or whether he will attempt to establish his claims by force'. Belgrade Radio replied that the Yugoslav Army was prepared for collaboration and agreement but not on a basis that insulted or humiliated her as an Allied country. There was a parade of force through the centre of the city one evening by Russian-built tanks of the Yugoslav Army. The Allies replied with its own demonstration of armour. The US joined in the public condemnation and it become clear that the Soviet Union was refusing to back its communist ally's territorial claims.

On 9 June, after two weeks of negotiation, an official statement was issued from London, Washington and Belgrade regarding the 'temporary military administration' of the area. Tito undertook to withdraw Yugoslav forces from the Italian territories which would come under Allied Military Government. The agreed deadline was 8am on 12 June.

The Yugoslav forces, including our neighbourhood machine-gunners, withdrew in a quiet and orderly fashion. On 12 June the Yugoslav flag was lowered from the Town Hall and replaced by the Stars and Stripes and the Union Jack. The Italian population hung out its flags and celebrated.

It was a relief to realise that Europe was finally at peace. Or so we imagined at the time. The Yugoslavs had a passionate hatred for the Germans and the Italians, and extended it to those of their compatriots who had collaborated with the enemy. But few could have foreseen how old hatreds would erupt in the current tragedy in the former Yugoslavia.

Between the troops of the western Allies and the Partisans in the city there was an undoubted measure of good will and respect. Looking back, it seems likely that Yugoslavia was less than totally committed to its claim on the city of Trieste, although solid in its determination to regain neighbouring parts of their country that had fallen under Italian rule – areas which were subsequently returned to them.

It may seem a trivial notion, but looking back I am sorry we did not approach the machine-gunners opposite our billet and ask them civilly to point their weapon in another direction. I think there was a fair chance they would have done so, at that time.

565 FSS
Loi Tek

Richard Kerr

565 Field Security Section, Intelligence Corps (India) entered Singapore with the leading elements of 5 Indian Division on 1st Septemer 1945. Within a very short period of time we had located members of the Singapore Police Force in a local internment camp (Simes Road?) one of whom proved to be Inspector Barrymore, a Canadian member of the local special branch. Inspector Barry was most anxious prior to his repatriation to brief us on special branch activities in order to enable someone in authority to pick up the threads in his absence. A particular point he made was that from his experience he fully expected problems to arise in the region from the activities of the Malayan Communist Party and in this connection he had placed an agent within this organisation prior to the war.

From what I can remember the French equivalent of special branch in French Indo China had contacted Inspector Barry concerning an Annamite agent of theirs that they had to get out of FIC. Apparently they had been using this agent and felt he could be of some use to us; his name was Loi Tek – or at least that was the name which he was subsequently known. It was decided to make use of Loi Tek within the MCP. Whether or not he was a known communist party member at that time I do not know. However, at the time Barry was interned by the Japanese he was controlling Loi Tek who occupied a minor position within the party. At the time of the Japanese invasion Barry destroyed all records of Loi Tek in order to protect his agent.

Barry was interned by the Japanese, Kempei Tai took over.

565 FSS now enter the scene. Our first task was to locate the Kempei Tai unit responsible for the Singapore area and these were found employed as coolies by our dock operating companies operating in Singapore Docks. It was my task to take these into custody, handing certain over to the PoW cage and retaining others for interrogation at our Balmoral Road headquarters. Interrogation proved a particular difficulty for us since none of us spoke or read Japanese. However, we were fortunate in locating a famous Japanese civilian by the name of Mamoru Shinosaki who had been involved in a spy trial in Singapore early in the war involving a Mrs Siddons whom he provided with monies to hold parties to which British officers and officials were invited. It was hoped that these people would discuss defence topics whilst in their cups and Shinosaki would circulate and endeavour to eavesdrop on their conversations. Shinosaki and Siddons had been on the black list when we entered Singapore and we had located both of them and found them most helpful to us. Shinosaki had been press attache to the Japanese Consul in Singapore pre-war but had become thoroughly disillusioned with the Japanese military during the occupation and was more than anxious to compensate for some of their excesses. He had a limited but useful knowledge of English and was thus useful for our interrogation purposes.

Using Shinosaki, we told the KT members that we required each of them to write out a full CV of their military career with details of places served, unit identifications, names of unit members, etc. We were particularly interested in this information to identify Japanese personnel responsible for war crimes which was another aspect of our duties. We impressed the KT members with the fact that we had a wide knowledge of events behind the Japanese lines from our agents and that we could check their stories. In the event this was not really necessary since the Japanese had never received security training and were more than anxious to glorify their military activity. However, we used Shinosaki to translate the CVs into English and we were then able to cross-check one CV against the others and build up a considerable amount of knowledge of the KT upon which with our detailed knowledge which had come, of course, from ourselves.

Resulting from these interrogations we came upon a report from the NCO of the Singapore KT who claimed to have been controlling an agent within the MCP. This was most interesting in view of the close liaison between Force 136 and the MCP and the information we had obtained from Barry. In brief the story was as follows:

The KT NCO (whose name I cannot recall) was allocated a role somewhat similar to that of Barry and was given Barry's clerk to assist him. The individual, anxious to ingratiate himself with his new boss, provided him with details of all Barry's contacts, one of which was Loi Tek. I doubt the clerk knew specifically the business between Barry and Loi Tek but was able to 'finger' the latter. Inevitably Loi Tek was located and picked up but proved uncooperative. Uncharacteristically for the usually brutal KT, this particular NCO used a more subtle methods of establishing a relationship with Loi Tek. He arranged to have himself locked in the cell with Loi Tek, the condition being that whatever the latter needed, even going to the toilet, was dependent upon his requesting the facility from the NCO. The NCO's patience was eventually rewarded and a relationship of mutual respect sprang up between the two men with the result that Loi Tek revealed that he had been working for Barry and was reporting on the activities of the MCP.

It was arranged that Loi Tek would continue this work under the KT. At this time Loi Tek was in a relatively minor position within the party and it was decided that it would be necessary for him to be advanced to a more senior position to give him greater control over matters. Accordingly whenever a secret meeting of the MCP officials was held the KT seemed to be aware and the majority were killed or captured and Loi Tek somehow managed to evade capture. Inevitably he rose to be secretary of the MCP as a result of these machinations, reporting to and directly controlled by the KT.

Whilst this aspect is crystal clear in my mind I cannot now be certain to what extent precisely the KT were informed of our plans but I do seem to recollect that they had a pretty good knowledge of what was going on and the part that the MCP was expected to play in the event of the British invasion of Malaya and Singapore. 5 Indian Division was supposed to land at Port Swettenham and it is my impression that the Japanese were aware of this.

My impression also is that Loi Tek was using both sides for his own ends although because of the methods he had employed to ensure his swift rise in the MCP he was more vulnerable to blackmail by the Japanese than he was susceptible to pressure from us.

The subsequent efforts of 565 FSS were greatly directed to locating Loi Tek who went to ground following the landing of British and Indian forces on the peninsula. I nearly caught him in Singapore in the winter of 1945/46. I had located a unit of the MCP in a house in Singapore located in the area of dense scrub and planned to take the members prisoner. Unfortunately the Military Government Commissioner of Police heard of the matter, took charge and using the Punjabi police units, made a frontal approach without securing the rear and Loi Tek escaped without difficulty, thus enabling him to organise many of the subsequent troubles of Malaya. It was all very frustrating and maddening.

574 FSS
Burma 1945

J.T. Lawrence

We embarked in an auxilliary cruiser and sailed with 355 Bde from Bombay. We rounded the peninsular and tied up at Madras. There was no disembarkation and the following day we were sailing again.

Capt. Frost called a 'prayer meeting' and opening a large envelope took out a number of files. Each showed the photograph of a wanted Indian, together with full description and address. All lived in streets in 'Millionaires Square' in Akyab.

Each of us had a file and we spent some time trying to memorise the details.

The Commandos were going to make a surprise attack on Akyab and we were to go in behind them, arrest our man and bring him out.

This caused some consternation and discussion in the cabin. It was my opinion that that this was too fantastic to happen – it didn't, we landed at Cox's Bazaar.

Making our way by river steamer down the Naf, we settled with 355 Bde at a tiny village called Nhila on the Teknaf Peninsular. From here Eddy and I travelled by khisti – a boat rowed by a man standing at the stern with a single oar – to remote villages where we talked to headmen, hoping for information regarding strangers to the village.

On one trip the khisti wallah pointed to the bank and said "Japani!"

This bothered us as at the speed of our man we were sitting ducks. Straining my eyes I made out what I thought were crouching figures. They turned out to be parts of the undercarriage of the plane.

The Commandos did their own thing and by and large managed without our help. Eventually they pulled out of Nhila and shortly afterwards we left too, finally settling again in Silchar in Assam, where we shared quarters with 588 FF Section whose FSO was Capt. Boyd.

Playing Bridge one night, I was studying the hand dummy had just laid down when he, dummy that is, knocking over his chair, made for the stairs. He was beaten to it by West whilst East made for the kitchen. Startled, I turned to look behind me, I found myself staring up the wide nostrils of a huge, black water-buffalo which had strayed into the mess.

Sgt Redding and I took leave from Silchar and making for Missoori in the hills, called in George Grasby who was very comfortable indeed in his Lucknow Section, 600 FS Section where he was CSM. Returning to Silchar we discovered that 574 FS Section had moved to Chittagong and that Capt. Frost had left the Section. Whilst in Silchar he had busied himself perfecting a means of getting bread to our detachments in a fresh condition.

The new FSO proved to be Capt. Percival Roberts, a one-time RSM of the regular army. He was quite different from Capt. Frost.

I went on detachment with a number of IORs to the railhead at Chandpur where the railway joins the Brahmapulra river ferry. It was fairly quiet here and I hired a Munshi to teach me some Bengali, without much success.

A letter from Frost told me that he was now FSO of 298 British FS Section where he was busy converting it into an Indian Composite Section like ours. In his letter he informed me that he was going to form his own private army. He had been successful in getting men he wanted and had taken Jackson, his former batman, from our Section, and also Havildar Abdul Hakina to be his Jemandar. His bombshell came towards the end of his letter. He had angled successfully to get George Grasby as his CSM and was looking forward to greeting him.

I knew George would be horrified at being winkled out of his cushy Lucknow Section where he was now running the show with no OC. George wrote to me expressing his feelings and saying that he had written to 'Joe' to inform him that he did not want to be CSM to Frost's FSO.

A movement order forced him to go however and wrote a letter to me later to say that after a week another movement order had sent him back to Lucknow. In George's words, "I stood not upon the order of my going – I went". Shortly after this I received a posting, it read, "Proceed to Imphal immediately to join 636 FS Section to be promoted CSM on arrival". I was unaware that 636 FS Section was Frosts. I was amazed when he picked me up at Manipur in his Jeep.

The Section was awaiting my arrival with the transport loaded and ready to roll. Capt. Frost had always wanted to be 'up with the fun' and his wish was now granted. We were now part of the 19th Division of the Fouteenth Army – The Dagger Division. The Dagger Division was moving rapidly south as unburied corpses witnessed. We camped in various jungle situations in bivouacs until at last we reached Toungoo where we settled in a timber-built house on stilts as all houses were constructed because of monsoon flooding.

The ever effervescent Capt. Frost immediately set about building his 'private army'. We attached to ourselves men of the former Burmese army, and a body of 'scouts' with an Anglo Burmese chap – George Savarese – as chief scout. There were out-of-work, timber-hauling elephants to be had and Frost got 12 of these to carry men and equipment to remote jungle detachments. Seven of the elephants carried the food for themselves and the other five elephants carried our rations. There was a concentration of Japs ten miles to the east of us. We offered a reward of 25 rupees for evidence of Japs having been killed by civilians.

Digging a 'bog' was a number one priority when we arrived at Toungoo. We avoided the roots of a large coconut palm and with the aid of woven bamboo created a creditable cover. The coconut palm had yielded to the prevailing wind over the years however, and we failed to notice that its head, carrying numerous coconuts, was directly over the bog. The first nut complete with massive husk demolished the roof and no doubt found sufficient nutrient to germinate in time. Until the new bog could be erected we had to use the original holding a zinc bowl over our heads. The second soon became infested with maggots and endeavouring to burn them out with petrol it was blown up. There was no fan but the distribution of the contents was much the same.

One night, sitting alone in the mess I was writing a letter home. There was complete silence apart from the scratching of my nib and the sound of distant bull-frogs. Suddenly I thought I heard a creak on the stairs which led down to the cook's area. The light of my tilly lamp fell on the top part of the bannister bat, beyond that was pitch darkness. After a pause and hearing nothing more I resumed writing but almost immediately heard another unmistakable creak. Looking towards the stairs I saw, to my horror, a yellow hand on the bannister. I was unarmed, my pistol being in my room, to reach it I had to cross the head of the stairway; I made my dash. At the head of the stairs my eye caught a glimpse of white clothing, I stopped and starred down. A Burmese man was followed by two women with another man bringing up the rear, "Can we come in please?" said the leading woman. Amazed, I let them in. The women took a chair each, the two men squatted on the floor. The woman who had spoken was tall and somewhat gaunt and very imposing. She introduced the other as her younger sister, Ma Sone, who was smaller, demure and very pretty. She then said they had walked a very long way, and asked if I had any bread for the men. I produced a loaf and knife and she cut it into chunks and gave it to them. Neither women ate any.

I was then informed that they were of the Chin tribe and also that the Japs were frequently near their village. They required arms and assured me they would kill Japs if they had rifles. At this point I decided to wake Capt. Frost, he was scantily dressed, but after a minute or two appeared in the mess and heard their request.

As I expected he jumped at the chance of another source of information and agreed to arm the villagers.

Some days later I had asked for toast and marmalade for tea and was eating it in the operations room when Capt. Frost strode in. He crossed to the window behind me and looking out said, "Finish your tea first and then come and have a look out ouf the window." Having swallowed the last bit I went to the window. In the compound below, a group of civilians were standing round the severed head of a Japanese soldier. They had carried it by means of a strip of bamboo fixed around a protruding vertebra and blood had run down over the face. In the hacking some of his cheek had been cut away revealing the teeth in a sort of a ghastly grin. We gave them their 25 rupees, and with difficulty I nudged it into a sack held open gingerly by a couple of our lads and we buried it in a corner of the compound.

One of our Burmese men became sick and our 'head-scout' informed me that he thought it was leprosy. I took the man and all his kit in the Jeep to the MO, who asked him to close his eyes and say when he felt something touch his face. He proceeded to touch the man with the point of his pen-nib. There were lumps on the man's face and he indicated that he felt the pen-nib, except when the lumps were touched. The MO declared that the man had modular leprosy. Some days later, to my horror, a similar nodule appeared on one of my fingers. I did nothing about it for a day or two, hoping it would go away. In the end I had to go to the MO. It was late in the afternoon, an orderly appeared and asked me why I had come at such an odd time, I replied that it was a confidential matter. Getting the wrong idea he said with some heat, "The MO don't look at them kind of things at tea time!"

"Tell him I've got leprosy," I said. Pretty soon the MO appeared.

"Who's the leper?" he asked.

"I am," I replied.

He looked at my finger and said, "Put your mind at rest immediately, this isn't leprosy, to catch leprosy you would have to live as closely as man and wife for about 20 years. What have you had your hand in?"

I rembered that I had set a prisoner to work line-washing round the base of our house and had got some of it on my hands. The MO said, "That's why you've got the blister."

I left elated.

Our Jemadar, Abdul Hakim, returning from a detachment visit asked me if I would return with him to try to get a woman to remove a curse which she had put on one of our men who was now convinced he was going to die.

I replied that he should remind the man that was a member of the Intelligence Corps and that this was the 20th Century. I declined to make a very long journey to the detachment. A few days later the man died and I was mightily relieved to learn that he had died of typhoid fever. We buried him near Toungoo.

Small numbers of Japs were crossing the Rangoon Road at night. Two of our NCOs rang into HQ one night having run out of petrol on the journey back from taking stores to a detachment there was nothing for it but for me to take petrol out to them. The hair on the back of my neck bristled from time to time on the way and when I made contact with them I let my feelings rip.

There was a small switch on the Chev dashboard which operated the petrol indicator. On switching it over to the other side it operated for the other tank. Behind the driver's seat was a large tap which had to be turned to change to the reserve tank. They had switced over on the dashboard but had not turned the tap behind the seat. They had one full tank.

Capt. Frost was eventually transferred to yet another section and was replaced by Capt. Burnett.

We then moved further south to Pegu. Before leaving Toungoo, Ma Sone presented me with a large and beautifully worked table cloth. I still have it.

Capt. Burnett himself left shortly after moving to Pegu, I never saw him again but continued to run the Section until my demob group was called. I left my Section with no FSO and no CSM, however the war was now over. The 'bomb' had done its work. We had for sometime been engaged on war crimes investigation and had heard some harrowing accounts.

One of my agents presented me with a watercolour painting of a typical Burmese village which, he said, had been painted by his son. I still have it.

Eventually I embarked at Rangoon and watched, with some emotion, as the Sheve Dragon Pagoda, gleaming with gold, faded at last into the mist as though it had all been a dream.

Many years later, I got George Grasley, Eddy Redding and Tom Frost together at my house. We met two or three times but on 9th November 1989, Tom Frost died aged 73 years. I attended his funeral and heard the vicar say that they had all known Tom Frost as the quiet man they often saw walking his dog in the lanes and had little known that he had been an army officer – they didn't know the half of it!

608 FSS

Jack H. Fry

I joined 608 FSS as the FSO on the 8th October 1945 from Burma. I soon realised that the appointment was special. I was under the direct orders of the Supreme Allied Commander, Lord Louis Mounbatten at the HQ in Kandy, Ceylon. At the initial interview he told me that as far as everone was concerned I was on leave awaiting posting, but I was to go to the Queens Hotel, Kandy where there was a party of high ranking Thai officers and their Liaison Officer, a Captain in the Guards, who were in Kandy to set up a Rice Trade Deal with the services. At this point I hadn't even met my Section. We were to set up a 24-hour surveillance with me on the inside and my Section taking it in shifts outside. The Rice Deal was a cover up, they were contacting civilian and service personnel and communist sympathisers. On leaving the interview I was given a bottle of whisky and a suggestion that it may help to keep me awake. Checking on the people they contacted and in turn checking out their contacts as required, kept us busy. My reports through GSO (I)b went directly to the UK via Foreign Office Cyphers; we actually got a 'Thank you' from the FO at the end. The other jobs we were given were of a similar nature.

On 27th November 1945 we embarked on Landing Craft Tank 350 for a 14-day voyage from Columbo to Singapore. The work in Singapore was routine, and after a few days I was summoned to

Mountbatten's HQ and told to take my section over the causeway into Malaya and they stressed that to be effective we must be the first Allied unit into Malaya from Singapore and remain one jump ahead of any other units, as the object was to get as much information as possible on War Crimes before troops swamped the area. Our other briefs were: to help the local population in as many ways as possible as they would be in very confused state: to pass on any relevant information to the civil powers, as and when they arrived: and to find the source of the gun running to the Ex-AJA, now the Chinese communists on the hills.

We entered Johare Bahru and set up shop in the late Sir David Galloway's residence, he had been murdered by the Japanese. It was a weird situation as there were no people of authority in the town, the Japanese had left and there was no one to replace them.; yet there seemed to be no violence or looting, not that there was anything to loot I suppose. The local police had, in the main, left with the Japanese for fear of reprisals by the local population.

As we were on our own, with no back up, we employed a Chinese family to cook and tend to us. One of our first tasks was to raid every photographic outlet in and around the town. These photographs were of Japanese soldiers and were to be used to help in the identification of war criminals. We then obtained the services of a Chinese/Malay interpreter, then set up a table and chairs and let it be known that we were here to help anyone who had been illtreated or in any way needed help.; it worked, and we attempted to sort out a mass of problems, marital, property, etc., and of course the all important war crimes, some of which were so horrific that it is hard to forgive that generation. After a few weeks there was a trickle of army officers sent in to start setting up the civil powers: the police for example, Capt. Terry Bray arrived and with the aid of our informers helped him to round up ex-policemen who had slipped away when the Japanese arrived, from then on we worked in very close co-operation with them.

It was brought home to us suddenly one day that we were a very small unit in what was in fact Japanese territory, although the war was over. Two Japanese officers complete with escort arrived on our doorstep, after the initial shock it turned out that they had a camp about 15km away and they wanted orders. I went out of the camp and it semed quite well organised and took the only action I could think of "Carry on and you will get your orders shortly". There was a bonus, the commandant gave us one of his soldiers to help with the chores around the Section HQ.

By this time Allied forces were establishing themselves in the north. We could now send reports and collect 'bribe money' from Kuala Lumpur. Troops were also now coming over the causeway, so according to our brief we moved up one state to Negri Sembelan and carried on as before. Our gun-running investigations were not bearing fruit, we knew who was sending the guns, a sikh wholesale merchant, but every time we passed the information to the police, who set up road blocks, it came to nothing. There was a leek and we took a long time to find it. Only very few people knew of the plans, and we could vouch for the informers. As we were in a peace situation all such raids had to be approved by the Judge Advocate General's Branch, and this was were the leak originated; a captain in that office had hidden from the Japanese when they overran the town, and he felt he owed them a debt. I showed him the error of his ways and the police were now more successful.

During this period we were on the communists' hit list, together with Malays and Chinese who in one way or another were helping the Allies. Sadly, Mr Lee, one of those on the list, a family man who helped us a lot and was a good friend, was murdered by the communists a few days after I left the Section on repatriation.

Reference the enclosed letter OB/26 'Personal Cleanliness and Sabotage', which may seem frivolous, but tends to indicate the way we worked. As a Section we were very much on our own with few comforts, hence the 'soap' in the letter. To the strict military mind we would have been classed as a shambles. I had 3½ years in an Infantry Regiment before being commissioned in 1942 into the IC (I). Regimental discipline was one thing but in our case respect would be more appropriate, as living and working together we were very much a team, we got the work done, each of us taking the next job – day or night.

FSS – Field Security Section

To OC 608 FSS OB/26

Subject/s: Personal Cleanliness and Sabotage.

1. *Soap enclosed.*

2. *Jerry carefully waited until we got down to Segamat, and then told me the following: Old man Moses (the engine driver not the prophet) took over an engine at Gemas at approx 0300 hrs 15 Apr '46. It was in fact the Mail Train.*

   During the journey, he was watching the fireman stoke the furnace, and saw a piece of putty, or something similar, on a shovelful of coal just before it was tipped into the firebox. He stopped the fireman and put the object aside. On arrival at Serembau he took it to the RTO Office and reported the matter. The CMP (I) were called, and it was decided that the object was HE. The CMP (I) took a statement from Moses, and are investigating.

   No further details, and the A/M is for your information. Perhaps you know already? If sabotage, its our job not CMP (I). Maybe its a Chases, Goose, Wild Mark III. HE. does not explode in a fire, it just burns, or does it?

   If you want me to do a Sherlock H. on it please communicate.

Segamat 16 April '46.

636 FSS
Burma 1945

Extracts from the War Diary by J. Lawrence

The author was Capt A.T.D. Frost

Tamu 1 May '45
Entered Burma at Tamu on 1 May '45, en route for Pynmana. Strength as shown on AFW 3008.

Meiktila 9 May '45
Arrived Meiktila on 9 May '45 where received verbal orders from GIII(Ib), 14th Army, that destination of this Section now altered to Toungoo.

Toungoo 12 May '45
Arrived Toungoo on 12 May '45. Under command of 4 Corps. Set up detachments : HQ Toungoo, dets: Yedasme, Pyu and Myhola.

Pegu 4 Jun '45
FSO contacts 4 Corps I(b) Staff.

Toungoo 6 Jun '45
Headman and Karen villagers from villages west of Toungoo called in to this HQ for conference. Co-operation promised from all in giving information concerning the enemy's moves west of Toungoo.

Toungoo 8 Jun '45
Scouts and agents organised to operate west of Toungoo. Personnel ex-Burma Rifles and locals being employed in this work.

Toungoo 10 Jun '45
Villagers bring in head of Japanese officer killed by them. Indentifications showing 54 Japanese Division in area west of Toungoo.

12 June '45
Field telephone laid 12 miles into jungle west of Toungoo to Pebin. (J. Lawrence here added that this was achieved with the help of small boys who shinned up trees to drape the line over branches) Extension to Kinziek for purpose quick transmission of information re enemy's moves west of Toungoo.

22 June '45
4 Japanese prisoners captured by our scouts and brought to this HQ for interrogation. 3 privates, one corporal.

30 June '45
8 Japanese soldiers known killed by our scout organisations west of Toungoo. Burma Police Inspector of Yedashe arrested in view of past co-operation with Japanese.

Toungoo 1–8 July '45
Attachment of 40 Burma Rifles as re-inforcements to the operation against small parties of Japanese escaping across West Route.

Toungoo 8–16 '45
21 Japanese PoWs and one comfort girl brought to this HQ by villagers. One Japanese killed by villager.

Toungoo 16–23 '45
Burma Rifles report in contact with enemy west of Toungoo. Final number Japanese killed in this engagement 35. Two of our civilian scouts killed.

Toungoo 23–31 '45
NTR.

1011 Field Security Reserve Detachment

F.P. Bowen

I joined 1011 FSRD from Buxton in May 1944, when it was with 1 Corps Field at The Bishops Road, in Hampstead. We went over to Arromanches in July 1944, and were stationed in the neighbourhood of Bayeux, and later with 1 Corps Field. Apart from reinforcing 30 Corps FSS around Caen for a period, and duty at Monty's Rear HQ from time to time (mainly refugee control), activity was practically nil until the Falaise Gap battle, when 1011 FSRD moved to the vicinity of Lisieux on the 23rd August 1944. It was then attached to 2 Canadian Corps, to provide the intelligence spear-head into Rouen, immediately following the Canadian armour clearing the axis of advance. The Seine was crossed at Elbeuf during the night 29/30 August, the bridge still being under enemy artillery fire.

1011 FSRD entered Rouen at first light on the 30th August after this rather exciting approach-march, made up of fits and starts and the sound of battle all around and the odd corpse along the highway. The various intelligence targets in Rouen were immediately investigated, and HQ set up in the Mairie, with the living quarters in some wooden huts near the cathedral.

A hectic period then followed, rounding up collaborators, Gestapo members and German Army administrative personnel who had been left behind, going through tons of captured documents – and trying to curb the enthusiasm of those French 'Resistance' members whose one idea seemed to be to undress and brand as many women as possible, on the flimsiest pretexts. On the evening of the 2nd September orders were received to get on the move – and this is the time the truck I shared with Sgt Wasley chose to break down. We were instructed to make for Brussels as soon as it had been repaired, 'R-Force' having disbanded. Having repaired the truck, we set off early on 3rd September, following Route 240 up as far as Amiens, were we stopped for lunch. As I knew the country very well, I suggested going via Lille to have a better through-road once we had got beyond Arras. About 10 miles before Lille, we saw anti-tank guns in the ditches on both sides of the road, pointing in the direction towards which we were going and the gunners, we thought, waved us a greeting. We entered Lille, widely acclaimed, just as it was getting dusk and not a little surprised at the enthusism displayed. We were shown the best hotel and told the room was 'on the town': it then dawned on us that we had just mistakenly 'liberated' Lille! We slept the sleep of the wicked to the sound of rifle and LMG fire all around, caused by the FFI mopping up the nests of German resistance, and awake in time to greet the first British armoured cars to enter the town!

We set off for Brussels early on 4th September, stopping for lunch at a café at the entrance to Hal. Fortunately, the truck had been parked in the garden, for while we were enjoying the Belgian publicans' hospitality, a tank with an ominous black cross lumbered across the road, fortunately free of 'soft-skinned' traffic at the time. Brussels had, of course, only been entered by the Guards' Armoured Division the previous night, and so many German pockets had just been left to their own devices, on either side of the main axis of advance. Brussels was, if anything, even wilder than Lille, and the Intelligence Task Force HQ was set up at the Grand Hotel for the first few extremely busy days, before moving out to Ave Louise, once 1 Corps Field had arrived and set itself up at the Cinquantenaire.

Working with the Dutch Resistance and later with 1017/21 FSRD and 7 FSS

Peter Scopes

Following D Day in June 1944, 9th Armd Div. was disbanded and I was posted to the Int. Corps Depot at Wentworth Woodhouse in July. In due course I was transferred to the SHAEF CI Pool and in September we landed as one of the first convoys in Ostend and proceeded to HQ Int. Corps (Field) in Brussels (OC: Major Sedgwick-Rough). I was posted to G (Int) HQ 8 Corps in Holland and in October I was awakened about 0500hrs by an orderly with a signal from Int. HQ in Brussels telling me to report with all my kit at 1200hrs outside the main Post Office in Eindhoven where two civilians wearing orange armbands would meet me and take me to Major Bornoff. I packed my kit and during breakfast I told a brother officer of my posting and he said: "Oh, my God – I hope your German is fluent and you speak the language without any trace of a foreign accent as Major B is the chap who infiltrates Allied agents behind the German lines". I proceeded to Eindhoven with fear and trepidation, was met by the two Dutchmen and taken to a house in the suburbs occupied by Major Bornoff. Much to my relief, he explained I was to take charge of the switchboard at the Nijmegen Central Power Station as unbeknown to the Germans, there was an underground telephone system which maintained contact with all their engineers, both north and south of the Rhine. There were 2 British sergeants and about 10 Dutch telephone operators or engineers at the Power Station and by a system of phone calls at specified times, we kept in touch with most of the engineers in the occupied part of Holland and they were able to provide valuable intelligence regarding German troop movements, etc. If information came through that a German unit or HQ had arrived in a certain town, this would be passed to our own HQ and the exact location would be bombed or shelled. The two senior Dutch engineers were Willem van Zanten and Jan van Gent – we got on very well and became good friends so we kept in touch after the War and I made several return trips to Nijmegen. Before the Battle of Arnhem, one of our jobs was to arrange for British agents to be ferried across the river so we would warn the nearest Dutch engineer in the German zone of the time and place of arrival and he would arrange for a member of the Dutch resistance to meet our agent. After the failure of 'Market Garden' there were numbers of British troops who evaded capture and were hidden by the local Dutch in their homes, farms, etc. In due course, by means of our telephone system, we made plans for groups of British soldiers to congregate on the river bank at a certain place and during the hours of darkness, boats were rowed across to bring them back to our lines.

This worked very well for a period but in due course by one means or another, the Germans discovered the existence of this underground telephone system and sadly, many of the brave Dutch engineers were arrested and executed, and the detachment at Nijmegen was disbanded.

My next posting was in October 1944 as OC 1017/21 FSRD at Weert (Limburg), Holland. Here we carried out normal security duties, vetting officials, tracking down collaborators, curfew checks, etc. One interesting case I had was when Corps HQ ordered me to arrest a Dutch Baron (M de W de W), and his Czech valet (K.L), as the Baron was suspected of collaboration with some senior German officers. We duly went to his Castle near Venlo and much to the Baron's indignation, we arrested him and his valet and took them to Corps HQ for interrogation. The young Czech valet was quickly dealt with and I was told to get rid of him! [As I was without a batman at the time, I asked K.L. if he would like to join my Section, explaining that in due course we would be advancing into Germany. As he had no roots in Holland and the Castle had been damaged by shelling, he accepted my offer. I quickly found him a Battledress and he (+ our Schnauzer dog 'Tommy') remained with me until I finally returned to England in 1946. As he spoke Dutch, German, Czech and some Russian and quickly learned English, and also how to drive a car and a Jeep, and was an excellent valet and groom, I found him very useful! When we finally parted, he returned to Holland, married his Dutch sweetheart, produced 14 children (one short of a Rugby team!) and we have kept in touch ever since. In due course, the Baron was released as he explained to the authorities that he was related to certain members of the German aristocracy serving with the Wehrmacht and his contacts with them were social and not political. (I maintained contact with the Baron until he died in Switzerland in 1975.) After the crossing of the River Maas early in 1945, my Sec was attached to HQ 3rd Brit. Inf Div and we advanced into Germany, crossing the Rhine at Rees, and finally arrived in Bremen when hostilities ceased in May. During the advance, we were briefed each night re the route we were to take and given the locations of all known Nazi HQs and offices in the towns we would pass through. We duly located them but needless to say, all the Nazi officials had fled, taking most of their records with them

After the cessation of hostilities in May 1945, my FSRD was disbanded and I was posted to Krefeld as OC 7 FSS. My Sec HQ was in the 'Arbeitsamt' (Labour Exchange), which was most convenient as there was a constant flow of civilians in and out so there were no problems for the steady stream of informers who came to 'Zimmer 7' to denounce various Nazi officials, etc, in return for cigarettes, coffee, etc I had a detachment in Munchen-Gladbach, and one in Viersen and one in Rheydt. Our work entailed tracking down Nazi officials who came within the 'AAC' (Automatic Arrest Category), apprehending them and placing them in the Allied Internment Camps for in-depth interrogation, checking all German nationals who applied for employment with Mil. Gov. (verifying their 'Fragebogens' etc).

There was a Transit Camp in Munchen-Gladbach and we had a checkpoint there to carry out security checks on both military and civilian personnel passing through. Our most interesting case started in December 1945 when a civilian sought out one of my Sgts and said he was Count Robert von Butler. He said he was a British subject as he had been born in London in 1916 of Austrian parents (he said his father was in the Diplomatic Corps) and he claimed he had been working for British Intelligence during the War in Germany and Norway. He said he was eventually captured by the Germans, tried for espionage and sentenced to death. Whilst in the condemned cell in Spandau prison in Berlin, he was liberated by the Russians and in due course he had managed to escape on a train travelling to the British zone. He explained that he was bi-lingual in English and German and in view of his intelligence experience, he would like to offer us his services in our counter-intelligence work, rather than return to the UK. He stayed with my Sec. for a time but I decided it was essential for him to return to the UK to establish his identity and obtain some valid papers. (My decision was hastened by the fact that one of my Sgts told me the Count had told the local German Housing Dept. that as he was working for British Intelligence, he demanded a private apartment in which to house himself, and his newly-acquired attractive blonde German 'secretary'! In addition, I was told he had sent for the best tailor and shirt-maker in the district and arranged for them to fit him out with a new wardrobe!) He was very reluctant to go but we got him on a plane and I advised a friend of mine in War Office (Int) of his estimated arrival time in London so he could be placed under supervision. I heard nothing more of him until April 1949 when I read in the English papers that a British officer had been court-martialled in Germany for the ill-treatment of of prisoners, and a Count Robert Butler-Brandenfels was brought into Court as a witness wearing a paper hood over his head. He claimed he had lived in England for ten months in 1946 until he had been arrested. I next read of him in the English and German papers in January 1954 when the 'Count' reported he had been kidnapped by Soviet agents in West Berlin in 1949 and was sentenced to 25 years in a Labour Camp in Siberia for espionage. However, he was released in 1954 with his fiancee and made his way to West Berlin.

I tried to make some enquiries about the 'Count' in 1954 but all my efforts came to naught – it was suggested that the man who reported to me in 1945 was either an imposter of the genuine Count, or maybe a double agent, and I was advised not to pursue my enquiries.

I had a very interesting and enjoyable time in Krefeld and district as we were engaged on all sorts of Field Security and counter intelligence duties. We had requisitioned a large private house with grounds on the outskirts of Krefeld, had German staff, we were allowed two private cars so I 'acquired' an Oldsmobile and a Sports Mercedes, and, of course, I still had K.L. as my batman and 'Tommy' my dog. However, in March 1946 my 'demob' number came up and I left my Section with some regret and returned to the UK to be demobilised.

Trieste Security Office 1948–49

Robert Andrew KCB

For a short spell in 1949 I was employed in one of the less publicised parts of the Trieste Security Office (TSO) 'A' Liaison Section, a joint unit staffed partly by ourselves and partly by the American CIC. In HQ BETFOR (British Element Trieste Force) was a door discreetly labelled 'Cabinet Office Historical Section' and behind it was all the apparatus of an extensive telephone tapping system kindly bequeathed to us. Locally enlisted staff did the actual listening, recorded anything they thought of interest and passed it to the US Special Agent or Int. Corps Sergeant on duty, who selected what should be passed on to higher authority for information or action. In fact, most of the information recorded was mere gossip (often of a salacious nature, which no doubt relieved the boredom of the listeners); but now and then there would be a discussion of plans for a political meeting; and just occasionally a snippet of information about arms smuggling. It was a strange world, made the more strange by the fact that we were on shift work; but it was a useful apprenticeship for the time 30 years later when I was Head of the Police Department in the Home Office and had to deal with telephone interception again in another context.

Those of us who worked in the Interrogation Pool also did some political reporting. The Italian elections of 1948 spilled over into Trieste and there were bitter divisions between the Christian Democrats on the one hand, supported by some small neo-Fascist parties, and the powerful Communist Party on the other. There were posters everywhere: I remember one graphic one of a communist jackboot poised over Istria, which seemed to echo the description of the future in Orwell's 1984 (which had just appeared) as 'a jackboot stamping on a human face, for ever'. Wall slogans, too, were highly developed, with the ubiquitous 'V Tito' which could quickly be converted by the opposition to 'M Tito' thus changing 'Viva' to 'Morte'. Large scale demonstrations were an important feature of the political scene: enormous crowds gathered in the Piazza Unita; and when processions of striking shipyard workers clashed with the police violence sometimes occurred. The VGPF (Venezia Giulia Police Force) were expert in the use of water cannon for crowd control, sometimes using purple dye to make it easier to pick up the participants later. We sat in our Jeeps in the background, in uniform with service revolvers in holsters, trying to estimate the size of the crowds and to distinguish what was being said over the loudspeakers. Some of our colleagues on the political side mingled with the crowd in civilian clothes which was more hazardous, and on one occasion Tom Snow was quite badly beaten up. One by-product of attending political rallies was that we learned the political songs. I remember fellow members of the Oxford University Labour Club being much impressed that I knew all the words of the 'Bandiera Rossa'.

We also tried our hands at a little low-level agent running, arranging to meet our agents (who were probably working for the other side as well) in discreet bars where we waited in civilian clothes but self-consciously aware of our short hair and generally non-Italianate appearance. For these occasions we were allowed an expense allowance which had to be accounted for monthly. I remember a colleague decided to test the effectiveness of the audit system. For the first month he submitted a bill for X lire 'for drinks with contacts', which was duly approved. For the second month it was Y lire 'for drinks with other members of the unit', and this too was approved. But the third claim for Z lire 'for solitary drinking' brought retribution.

From time to time there were moments of greater drama. The

VGPF acted as our executive arm in arresting people for questioning and the 'three o'clock knock on the door' was the accepted practice. Sometimes one of us went along to observe. For these and other occasions we were issued with a special Field Security pass, of which we were rather proud, entitling the holder 'to be at any place, at any time, to wear civilian clothes and to carry a civilian weapon'. The VGPF were not very gentle in their methods and to see them arresting a man in the middle of the night and ransacking his apartment while his wife and children stood by in their night clothes crying was a distasteful sight. (Wasn't this one of the features of the Nazi police state we had just fought a war against?)

Perhaps the greatest anti-climax came in the summer of 1948 when Tito broke with the Cominform. When news of the split first came through no-one knew what its implications would be and there was much speculation and excitement in TSO. Would the Red Army move into Yugoslavia to restore Moscow's authority? How would the local communist party in Trieste react? We were despatched in our Jeeps to the border checkpoints, not knowing whether to expect to see Soviet tanks or a flow of refugees. In fact, at my particular checkpoint nothing appeared all day, except for an old woman in black with a donkey and some goats.

Of course, life in Trieste was not all field security duties. Living in a European city was a new experience. After the austerity of wartime Britain it seemed a place of luxury, with well-stocked shops; we had silk shirts made to measure and bought nylon stockings to take home as presents. There were pavement cafes among the trees of the Viale XX Settembre serving delicious ice cream in unimagined variety; and restaurants in which we learned to eat spaghetti and were introduced to dishes like the pizza (then unknown in England) and scampi – not to mention inexpensive wines. There was opera in the courtyard of the Castello San Giusto in the summer and at the Politeama Rossetti in the winter. Apart from a few occasions in winter when the 'bora' blew, and ropes had to be stretched across the streets to help people cross, the weather was generally good and warmer than most of us had been accustomed to. From 1 April to 1 October we exchanged battle dress for khaki drill and at the weekends took the tram to the bathing beach at Barcola. And there were indeed Italian girls of remarkable beauty, some of whom were only too anxious to acquire a British passport. An Italian girl friend was considered one of the best ways of improving one's knowledge of the spoken language. On returning to England I found the female undergraduates of Oxford distinctly dowdy by comparison!

One of the advantages of Trieste was that, although it did not itself rate many stars in the Michelin Guide, it was readily accessible to a good many agreeable places. We could get to Venice for the weekend and further afield for local leave. With John Perkins I went to Rome and Naples (flying, for the first time, in a Dakota), to the Italian Lakes at Easter and to Cortine d'Ampezzo in the Dolomites (not yet famous as the site of the Winter Olympics) to learn to ski at the Army Leave Centre. It was the next best thing to the Grand Tour of the 18th century – and at the taxpayers' expense. All in all it was an interesting and enjoyable way of spending two years and I think that, as a result of it, most of us went up to university a good deal more mature than we would otherwise have been.

Some Random Recollections

W. Sedgwick-Rough

The somewhat mysterious signal signed 'DIRMILINT', containing a summons to attend the Field Security Course at Sheerness early in 1940, gave no indication of the challenges, excitements, and hazards of the next six years.

At the end of the course a working knowledge had been gained of the Security Systems evolved in conditions of Trench Warfare 22 years previously: and the German Blitzkrieg in Western Europe followed by the retreat to Dunkirk rendered it totally irrelevant.

Command of a Field Security Section in East Anglia, and later in Gloucestershire, provided the opportunity to begin to consider the role FS should play, in the circumstances likely to evolve – only the barest outlines of which could then be perceived.

Command of the Advanced Training Company at the Intelligence Corps Depot at Winchester the following year, informed by reports filtering back from Abyssinia and the Western Desert, provided further enlightenment: and this, together with the rudiments of military training – including motorcycle training, (which was frequently lethal) – was inculcated into the 180 new Intelligence Corps Members who were breathed into and out of the Company each fortnight.

This rapid expansion of the Corps with a bewilderingly versatile and varied array of talent was so urgent that finer points had to be ignored: for instance, a fluent Japanese speaker (later a pearl of great price) would find himself in Iceland, while a fluent Persian speaker would be posted to Northern Nigeria.

A staff appointment as GSO 3(I)b, and attendance at the War Intelligence Course in Derbyshire, was an admirable interlude before becoming Field Security Officer of 11th Ammoured Division, then in Newmarket.

The carpet-slippered life of a very somnolent – but elegant (and precious) – Field Security Section located in Queen's College, Cambridge, came rapidly to an end with a move to a stable in Moulton Paddocks, Newmarket: and the Field Security Section found that the day started with a long run in the snow on Newmarket Heath, stripped to the waist, followed by strengthening exercises, including heaving heavy logs and tree trunks about, which produced a sharp and physically hardened Field Security Section in a very short time.

Exercises with the Division, and study of our intended active service role in North Africa, in particular the Capture of Kairouan, assisted the equally sharp minds of the members of the Section to clarify the purposes which we should serve: and three-day exercises on the Yorkshire Wolds, living out in thick snow in mid-winter, helped concentrate our minds wonderfully on the techniques of survival to be fit to fight in adverse circumstances. Shortly before the Division was due to embark for North Africa, a lecture was given by myself, in the then normal way, to one of the Division's Infantry Battalions concerning Security. Afterwards, over a drink in the Officers' Mess, an officer said how very interested he was in the lecture since he had an aunt living in Tangier and had written to her to say that he expected to be seeing her shortly. On my return to Divisional Headquarters, the right of direct access granted by the Divisional Commander to his Field Security Officer was utilised, and his speed of reaction was both impressive and remarkable, resulting in the visit next day of a senior and extremely formidable member of MI5, whose modus operandi, if not exactly a joy to behold, was nevertheless among the most powerful performances I have witnessed in (what is now) quite a long life.

The personnel of the Divisional Field Security Section were by now hard, fit, confident in their ability to perform the various tasks pertaining to their function, and with a tremendous esprit de corps, since they were regarded not only by the Divisional Commander and his staff, but throughout the Division as a whole, as an essential and worthwhile component of the whole formation which, however, after eight false starts, never reached North Africa, since it was finally decided that the requirement was for an Infantry and not an Armoured Division.

This change of plan resulted for myself also in a change of role, namely to the Command of Intelligence Corps (Field), intended to become the Headquarters of the Intelligence Corps in 21st Army Group; and at the end of 1943 it was a reasonable guess that Operation Overlord, and the Invasion of Europe, could not be much more than six months away. There was much to be done in the location in Highgate where the expansion to a strength of around 1,000 was to take place. The accommodation was drab, especially the dining room: accordingly an excellent, but somewhat avant-garde artist, Corporal Peter Samuelson, was persuaded to paint the necessary murals on the walls. When completed these turned out to be subtly and deliciously pornographic, which resulted in overwhelming customer satisfaction, and (who knows?) better appetites.

Training proceeded in like manner, for each unit or draft, to that applicable in 11th Armoured Division, with one important difference: proximity to and frequently long absence from the capital city resulted in general monumental hangovers, which could only be dispelled with much suffering, but usually had worn off by the end of a three-mile run led by myself, all stripped to the waist, followed by 30 minutes hardening exercises.

The documentation concerning country-study and related matters, received from 21st Army Group, was excellent and enabled a detailed uniform doctrine, and the manner and means of implementing it, to be firmly established. The standard of training and the quality of Intelligence Corps personnel was a source of satisfaction, and even comfort, in the face of the unknown challenge with which we were all confronted. Field Security Units of other nations were also trained, notably United States CIC Personnel, Poles, Canadians, Free French and the like.

A number of the NCOs in the 11th Armoured Division Field Security Section had by now been commissioned and were themselves commanding Field Security Sections within Intelligence Corps (Field). A card index system of details of every Intelligence Corps member was established containing inter alia details of any part of

Western Europe with which they were familiar, their linguistic abilities (which were tested), and a photograph. The first of these latter two frequently produced a substantial difference between claim and reality, and the second resulted in the Commanding Officer being credited with a far better memory than he possessed. Another advantage was clearly that any members with local knowledge of an area could be directed to where they could serve most advantageously, with a ready made array of local contacts.

Finally the great day for the move to Normandy came, which of course called for some prior celebration: this took the form of a friendly confrontation between the Officers' and Sergeants' Messes with plenty of thunder-flashes on either side. This produced panic reports in High Places that the Germans were bombarding London with a new type of missile – but before the true state of affairs could be ascertained we were gone.

Before leaving Highgate we had acquired a Code Breaking Section and an Intelligence Laboratory, which later was landed on the wrong beach in Normandy and took some finding. Apart from that we all arrived on the Beachhead intact, although the Commanding Officer's arrival was in an unceremonious fashion. Lance Corporal Wright, never a strong motorcyclist, contrived to drop his vehicle into the sea, after which it would not start. It was obviously neither the time nor the place to hang about so, with the Commanding Officer riding it, Sergeants Alf Langdon and Ginger Farrow towed the vehicle to its appointed station, along a route which could reasonably be described as noisy, hazardous, and very uncomfortable. But, once ensconced comfortably with a bottle of gin, in his slit trench, the Commanding Officer was better able to begin to assess the situation, which can best be epitomised by the reply of one of the FSOs: "Absolutely marvellous: everything you have told us was right and in fact it is just like another of your bloody exercises'. The only discordant note was the daily visit from the two scientists enquiring plaintively after their Intelligence Laboratory (eventually found intact and subsequently set up in the Hydropathic Institute in Brussels).

Reports and visitors filtering back made possible the refinement of the common doctrine in the light of battle experience: and this enabled the additional Field Security Sections and individual NCOs arriving from England to have the benefit of such experience. Such additional manpower brought the strength up to 1500 approximately.

The move to Brussels was preceded by the formation of an ad hoc Intelligence Corps Battle Group, with a strength of 155, and with the task of securing all the Intelligence Targets in the City: this was accomplished. Intelligence Corps (Field) was established in Brussels in eight houses in the Avenue de Tervueren, close to the Cinquentennaire. By then, Field Security Sections were operating in France, Belgium, Holland – and later, of course – Germany. Allocated to us were Security Liaison Missions from France, Belgium, Holland and India, whose task was to collect and return their nationals who had fought alongside the Germans. (We heard later that President Nehru had greeted as heroes the Indian contingents on arrival in the subcontinent and had released them immediately, much to the chagrin no doubt of the retired Indian Army Colonel who led the Mission, and who had worked hard and long to find them.)

The normal work of processing and deploying, or redeploying, Field Security units and personnel and training or rebriefing them, together with all the normal administrative functions (not least with regard to that famous Intelligence Laboratory, which we understood was beavering away at . . . no one knew precisely what), was interspersed by visits to Field and Port Security Sections in action.

The striking and almost eerie emptiness of the Port of Antwerp the day before it reopened is an abiding memory.

The final score of agents caught ran into hundreds: one particularly favoured and aquatic infiltration route proved so productive for us that it became known as the Biesbosch Regatta.

Other delights of a miscellaneous nature presented themselves from time to time: one such was what the Germans called a Grossfahndung. Brussels was full of German and other deserters, some German agents, and various other strange elements who had no business there, but were, in sum, undesirable. It was thus decided to appehend them: a curfew was imposed; and Intelligence Corps personnel with Military Police co-operation set about the task.

The results were striking and included surprises such as a senior lady in the Women's Forces who was very much in the wrong place at the wrong time and, in old-fashioned terms, doing the wrong thing. Belgian friends who had been really active in the Resistance thought that our performance was much superior to that of our opposite numbers, the Geheime Feld Polizei: and a few days after our Grossfahndung the barber cutting my hair remarked 'on m'a dit que la Gestapo Anglaise est beaucoup plus terrible que la Gestapo Allemande': a backhanded compliment perhaps . . .

By this time Field Security Sections had assumed a remarkable buccaneering air – but highly individual to each Section. Motorcycles had been discarded, since none of us fancied carrying, on such a vehicle, a large SS man or equivalent, to summary justice or worse.

A remarkable variety of captured vehicles including amphibious Volkswagens had been acquired, plus variegated clothing, and frequently the odd Russian who became chief cook and bottlewasher, which he invariably preferred to the forced-labour building of the German Atlantic Defence, which had previously been his lot.

Towards the end of 1944 it was felt that, in the interest of public relations, a ball should be given in the Officers' Mess, for which invitations were dispatched to the many kind and distinguished Belgians who had tolerated us, as well as to our own Officers and representatives from 21st Army Group including the BGS (I) Brigadier Williams. Before this could take place, however, Von Rundstedt started his Ardennes Offensive, which resulted among other things in the United States 1st Army being placed by General Eisenhower under command of 21st Army Group.

One of the problems our United States' comrades were faced with was the infiltration through their lines of English-speaking Germans in United States' uniforms: accordingly Intelligence Corps (Field) was ordered to produce 110 of their best German speaking personnel to assist them. Within three days another ad hoc Battle Group had been withdrawn from their previous locations, constituted, briefed, totally re-equipped and delivered to US 1st Army. It contained a sprinkling of the original 11th Armoured Division Field Security Personnel, now nearly all commissioned. We received a special letter of commendation signed by General Omar Bradley.

While the ball was only a poor affair compared to that given by the Duchess of Richmond 139 years before, prior to the Battle of Waterloo, it had nevertheless a certain evocative flavour: the principal difference being that in our case the Battle of the Bulge was going on at the time, and people wondered whether the tanks passing by were German or our own. I hope that Brigadier Williams, still I believe in his City of Dreaming Spires, will not mind my recalling that at four in the morning he was playing the drums in the Band, in a condition I can venture with due respect to describe as happy and carefree: but I suspect he already knew that the Battle was won, while to us this was not in any way apparent. His musical performance caused a certain disquiet among the 'Naphthalines', as those Belgian and other Allied Officers were called, whose much be-ribboned tunics had been preserved for the last five years in moth-balls: but this disquiet was dispelled by the outcome of the battle.

Life returned rapidly to normal in the bitterly cold winter of 1944 with occasional special features. One such was a visit by a very senior general which necessitated a momentary return to spit and polish. Once the situation was explained to them the Field Security Sections went off and voluntarily practised and perfected their parade ground drill: they requested one combined parade to become again accustomed to the Commanding Officer's parade ground voice. On the day they were magnificent and attracted lavish praise – after which we all had a good laugh and a drink although one of our buildings was hit by a rocket at lunch-time, incurring a number of casualties.

As the campaign progressed new requirements arose – for example a Field Security Section for Denmark as that country approached liberation, composed naturally of Danish and Norwegian speakers. A Field Security Section composed of Russian speakers was provided for the Potsdam Conference: this was commanded by Captain Alan Eade who had been one of the original 11th Armoured Division Field Security Section of three years previously. We were thus able to sum up and judge our Russian opposite numbers.

By this time the whole scene in Western Europe had changed from a war-time footing. Representatives of strange bodies arrived; for example the Catholic Women's League and the CVWW (Council of Voluntary Women's War-Workers): with what – if any justification cannot now be recalled but they came to be known as the Careful but Very Wicked Women.

By mid-1945 it was clear that barring accidents the Allies would win the war and it was therefore time to give thought to the stay-behind organisation which the Germans were expected to leave, known as Werewolf. Its elimination was to fall to our lot for which purpose a close co-operation was established with 'B' Squadron SAS. In fact this threat failed to materialise and the few sabotage dumps established were found quite easily and destroyed.

Once the war was over there was still, for a while, work to be done, apart from collecting up persons such as Himmler, Ribbentrop and

Alfred Rosenberg, together with whole categories of German Personnel such as Abwehr and Gestapo members, Waffen-SSs and the like. But by the time of the move of Intelligence Corps (Field) to Bad Salzuflen, it was clear that our motto should be changed from 'No advance without Security' to 'The Party's over my friend' – or equivalent.

In Berlin we found the German opposite number of our intelligence laboratory scientists. A highly satisfactory technical meeting between the two parties was arranged at which details were compared as to how they had been working against each other in the past, before combining their talents as to how they would, together, out-wit and out-fox their Russian opposite numbers.

Before leaving Brussels we had held for a time a KRIPO member Hermann Kamau, who was the first person to testify that he had seen the bodies of Hitler and Eva Braun, with petrol poured on them, being burned in the Berlin Bunker. We had also held for a time the wife of Lord Haw Haw for interrogation: a vitriolic and very racy lady!

In Bad Salzuflen 'demobitis' raged; the Control Commission arrived to take over (they were a shambles by our standards) and our task was concluded. All that remained was to bat out time until the moment came to undergo the rigours of civilian life in post-war England which for me was in January 1946.

One final vignette of this period – from Magdeburg, which our troops had captured, but which was to be handed over to the Russians: the Field Security Section became rather bored with collecting up those on the 'wanted lists' but someone, during a heavy and hilarious evening, had the bright idea of putting an advertisement in the local paper ordering all persons in the required categories to report at 8am on a particular day, to the Field Security Office and then all was forgotten about it. On the appointed day a long queue composed of the persons required formed outside the Field Security Office at 7.45am much to the surprise of the Duty NCO. An interesting application, perhaps, of the old German saying: Dienen meint Gehorchen. Some little time afterwards all our troops withdrew from Magdeburg, very much to the surprise, and no doubt delight, of the Russians and to the horror of the remaining Germans.

Thus we all went either home or at least our separate ways, in the knowledge that at any rate so far as the war against Germany was concerned our task had been accomplished satisfactorily.